PREDICTING STORM SURGES:
Chaos, Computational Intelligence, Data Assimilation, Ensembles

DISSERTATION

Submitted in fulfillment of the requirements of
the Board for the Doctorate of Delft University of Technology
and the Academic Board of the UNESCO-IHE Institute for Water Education
for the degree of **Doctor** to be defended in public
on Tuesday, 6 December 2011 at 15:00 hours
in Delft, The Netherlands

by

Michael Baskara Laksana Adi SIEK

B.Sc. in Mathematics, Airlangga University, Indonesia
B.Com. in Information Management, STIKOM Surabaya, Indonesia
M.Sc. in Hydroinformatics, UNESCO-IHE Institute for Water Education,
The Netherlands
born in Jember, Indonesia

This dissertation has been approved by the supervisor
Prof. dr. D. P. Solomatine

Composition of Doctoral Committee:

Chairman	Rector Magnificus Delft University of Technology
Vice-chairman	Rector UNESCO-IHE Institute for Water Education
Prof. dr. D.P. Solomatine	Delft University of Technology / UNESCO-IHE (supervisor)
Prof. dr. A. Heemink	Delft University of Technology
Prof. dr. D. Roelvink	Delft University of Technology / UNESCO-IHE
Prof. dr. H. Kantz	Max Planck Institute for the Physics of Complex Systems, Germany
Prof. dr. M.C. Deo	Indian Institute of Technology, India
Dr. M. Verlaan	Delft University of Technology / Deltares
Prof. dr. A.V. Metrikine	Delft University of Technology, reserve member

Published by CRC Press / Balkema Publishers, a member of Taylor & Francis Group.
www.crcpress.com – www.balkema.nl – www.taylorandfrancisgroup.com

ISBN: 978-0-415-62102-1 (Taylor & Francis Group)

Front cover: Lorenz63 attractor with parameters (a=1.166; b=6.242; c=7.184; dT=0.224)

This thesis is dedicated to the memory of my father:
Maxim Sila Chayka – 薛天才

Summary

Over the past centuries, a number of severe coastal floods due to storm surge have occurred and had destructive consequences in many places in the world. The physical mechanism leading to coastal floods is now well understood. The severity of the storm surge depends primarily on meteorological forces, such as air pressure difference, wind speed and direction. The meteorological conditions are affected by the path and the velocity of the depression systems moving across the sea. When winds push water towards the coast, they tend to accrue into what is commonly referred to as storm surge. If a particular high surge occurs together with high tides, both effects amplify and can result in increased sea water level and serious flood in coastal areas.

Accurate predictions of storm surge are of importance in many coastal areas. Particularly in the Netherlands, reliable storm surge models are of great importance since the large areas of the land lie below sea level and the storm surge often occurs in the North Sea. Defenses against floods by the sea have been systematically improved, such as constructing storm surge barrier designed for 10,000 years return time period of extreme storm and building more sophisticated model for predicting storm surge. The storm surge predictions and warnings are made by the Dutch storm surge warning service (SVSD) in close cooperation with the Royal Dutch Meteorological Institute (KNMI). The model predictions for at least 6 hours ahead are required for proper closure of the movable storm surge barriers. These predictions are based on a numerical hydrodynamic model, the Dutch Continental Shelf Model (DCSM) which receives the meteorological predictions from High-Resolution Limited Area Model (HiRLAM) as driving forces. A data assimilation technique based on Ensemble Kalman filtering has been added to this system to improve the prediction accuracy by assimilating the recent observations from tidal gauges. The other significant improvements that have been brought into the model, include: refining computational grids, calibrating the model, using a better numerical scheme and implementing data assimilation techniques (3D/4DVar and Kalman filter). Note that the prediction accuracy of a storm surge model based on Navier-Stokes equations, like DCSM mainly depends on the accuracy of meteorological predictions from the weather model (i.e. HiRLAM).

The model mentioned belongs to the class of process models (also called physically-based or numerical models). The present study focuses on a quite different modeling paradigm, known as data-driven modeling (DDM), a modeling technique which primarily uses the analysis of the data characterizing the underlying system. The model is mainly defined on the basis of connections between system state variables (input, internal and output variables) with only a limited knowledge of the details about the physical behavior of the system. The approaches in data-driven modeling generally originate in statistical methods and artificial intelligence. Several popular models in DDM include: artificial neural network (ANN), instance based learning, model tree, Bayesian learning, committee machine, fuzzy rule based system and genetic programming.

Yet another approach in data driven modeling is based on the use of the methods of nonlinear dynamics and chaos theory which are typically applied for modeling complex dynamical systems. These methods have been effectively enhanced with considerable emergent research since Edward Lorenz made a discovery in 1963 during his experiments with a simplified atmospheric model. He explored the sensitivity to initial conditions that leads to chaos theory. The meaning of it is that a dynamical system derived from differential equations can exhibit chaos, which has a characteristic of the exponential divergence of the model outputs if the initial conditions are slightly perturbed. Subsequently, a number of researchers and scientists investigated and modeled many kinds of natural phenomena and discovered that many have chaotic behavior, whereas previously these natural systems were believed to act randomly. The dynamical systems that are characterized by deterministic chaos are predictable. In this work, we have the luxury of having very large data sets characterizing the dynamical system, in this case storm surge, which gives us the possibility to show that it is chaotic (i.e. without direct us of differential equations describing this system), and that the predictive data-driven model can be built.

The main objectives of this work to is to build a more accurate chaotic (data-driven) model that can serve as a complementary model to the existing operational storm surge models for the North Sea region. More specifically, the objectives are: to analyze deficiencies of the existing methods and enhance techniques for building predictive chaotic models, incorporate data assimilation methods into chaotic models, to develop and test multi-model ensemble approach to combining various predictive models. Main parts of the methodology are nonlinear dynamics and chaos theory, data-driven modeling, process based modeling, data assimilation, optimization and ensemble methods. In general, we can classify this study to be belonging to the area of hydroinformatics. The main case study for this work is surge prediction at Hoek van Holland tidal station in the North Sea. We also tested some approaches to optimization of chaotic models using the data on surge at San Juan tidal station (Puerto Rico) in the Carribbean Sea.

The initial experiments of building a univariate chaotic model for predicting storm surges in the North Sea has been conducted by Solomatine *et. al.* (2000). In the PhD study of Velickov (2004) this approach was further developed and extended the predictive chaotic model (PCM) into a multivariate model, which can include other variables, such as wind and air pressure. The nonlinear time series analysis of the observed surge data indicates that the storm surge dynamics along the Dutch coast can be characterized as deterministic chaos. Chaotic behavior in the storm surge dynamics can be due to the fact that this dynamical system is the result of complex interactions between different forces or dynamical systems, such as atmospheric dynamics and wind-wave-tide interactions. The presence of deterministic chaos and positive largest Lyapunov exponent implies the possibility for prediction. However, predictability of any model including predictive chaotic model has some limits. The properties of the sensitivity to initial condition and the existence of bifurcations can be some reasons associated with exponentially decreasing prediction accuracy of chaotic model after a certain time of prediction horizon. Nevertheless, the short and medium-term predictions of this model are generally quite accurate.

In constructing a predictive chaotic model, the observed time series of a dynamical system needs to be reconstructed and embedded in a sufficient m-dimensional phase space with time-delayed coordinates/manifold. This reconstruction preserves the properties of the dynamical system which do not change under smooth coordinate adjustment, but it does not maintain the geometric shape of structures in phase space. The proper values of time delay and embedding dimension can be estimated by means of several nonlinear analysis tools (e.g. first minimum mutual information and correlation dimension, respectively), or optimization methods. Given the proper dimension and time-delay of a phase space, the attractor of a dynamical dynamics should be unfolded and subsequently the smoothed trajectories are obtained. Predictions in chaotic model can be made by two ways: using global or local modeling.

In global modeling, the whole dynamical behavior of the system as described in phase space is characterized and predicted by single global model. In contrast, the local modeling allows for characterizing the dynamical behavior locally by a number of local models and the options on determining predictive local models are more flexible. The local models are constructed by the dynamical neighbors found in the phase space. Several available data-driven techniques (i.e. linear or non-linear regression methods like ANN) can be utilized as local models. Nonetheless, the flexibility of local models presents a challenge of selecting the best searching technique for finding true dynamical neighbors and choosing the suitable number of dynamical neighbors used for building the predictive local models. The true neighbors here refer to neighbors that have the similar dynamical characteristics or properties (i.e. similar type of storm development) to the reference or actual points in phase space. In this research, Euclidean distance method is employed for searching dynamical neighbors. The searching algorithm used earlier was not very selective and was sometimes finding these dynamical neighbors that do not have similar dynamical characteristics, so that they were wrongly treated as neighbors by the algorithm. In this work this issue received special attention, and a new searching technique, so-called trajectory based method, is introduced for avoiding the false neighbors.

The methods and some software components developed in earlier work have been integrated, tested on the new data and considerably improved in a number of directions. Innovation brought by this work is in the following.

The new algorithm for identifying the true neighbors has been developed and tested. It is named the trajectory based method and arises from the main idea that finding true neighbors does not only depend on the distance between two points in the m-dimensional phase space, but also the distance of the two different trajectories (sequences of points in phase space) partly formed by these two points. The neighbors are obtained by searching for trajectories which are nearest distance and similar direction to the actual trajectory (a trajectory formed by the reference or actual point in phase space). Other methods for avoiding false neighbors, such as using multi-step prediction technique and neighbor distance cut-off method, are proposed in this work as well.

Identification of suitable embedding dimension is the most discussed topic in the community of nonlinear dynamics and chaos theory. For example, a correlation dimension is a widely used method for estimating embedding dimension. This estimator requests for large-size of time series data to provide good embedding dimension estimation. In this research, the result of correlation dimension is compared with false nearest neighbors, Cao's method, Kaplan-Yorke or Lyapunov dimension and performance-based optimizations. The techniques in computational intelligence, such as grid search, genetic algorithm (GA) and adaptive cluster covering optimization (ACCO) are utilized in this work for performance-based optimizations.

Several other innovative developments of the predictive chaotic model have been made including phase space dimensionality reduction, building chaotic model from incomplete time series and correcting phase prediction errors. The nonlinear analysis of time series from a dynamical system may suggest the high-dimensional phase space reconstruction. A principal component analysis (PCA) technique is utilized for reducing the phase space dimension into a lower one by preserving important information (principal components) in high-dimensional phase space (i.e. distance information) into lower-dimensional phase space. Another benefit of applying PCA here is that it can remove the noises that may exist in the data. A procedure for building a predictive chaotic model from incomplete time series is crucially required in the view of the fact that measurement instruments and data transmission do not always work in real-operations. The possibility of missing some data should be addressed when building a model. Several imputing algorithms, such as weighted sum of linear interpolation, Bayesian PCA and cubic spline interpolation are proposed to resolve this issue. An approach of building a model for characterizing the phase error dynamics is proposed for correcting phase prediction error in the chaotic model. Two types of models are used as error predictors (predictive chaotic model and ANN), and they are able to identify and predict the dynamical behavior of the phase error generated by a standard chaotic model.

A number of approaches have been tested in order to address the issues related to sensitivity to initial conditions and the limitation of predictability of any model, including a predictive chaotic model. Resolving the issue of sensitivity to initial condition by finding the precise and exact initial conditions is not an option. The possibility to resolve this issue is to introduce data assimilation scheme into the predictive chaotic model. A Nonlinear Autoregressive with Exogenous Inputs (NARX) neural network has been implemented as a nearly real-time data assimilation technique for assimilating the new observed data into the predictive chaotic model. This technique can effectively correct the low accuracy of predictions after a certain time of horizon, and subsequently extend the predictability of the chaotic model.

Yet another innovation is using multi-model ensemble predictions: they have been viewed as an effective way to improve the prediction performance (based on bias-variance decomposition) over what the single models can provide. It is often worthwhile to seek a combination of several prediction models rather than to select only the best one among them, which might be only marginally the best. Multi-model ensemble predictions using dynamic averaging and dynamic neural network model are introduced for combining the heterogeneous types of predictive chaotic models. A dynamic averaging method is introduced – a combination of model selection and model combination approaches based on the model performances over certain time of predictions. The other technique uses one type of dynamic neural networks, so-called Focused Time-Delayed Neural Network (FTDNN). Several predictions from different types of predictive chaotic models are selected and further combined by these two techniques in order to obtain more accurate and reliable predictions. In terms of a high-dimensional chaotic system, it means the ensemble of all future trajectories in phase space, estimated by the heterogeneous individual models.

A number of improved methods of building predictive chaotic models has been implemented and tested. The results showed the increased predictability and performance of the initial predictive chaotic model: PCM is 63% more accurate than ANN model; univariate-PCM with PCA can increase the accuracy by 118% compared to multivariate ANN; 94% performance increase is achieved by using PCM error corrector; reduced accuracy by as low as-8% is given by cubic spline interpolation in case of 30% missing values, trajectory based method can better find the true neighbors resulting in predictability improvement by 185%; adaptive cluster covering optimization method (ACCO) appeared to be the most efficient optimization technique for predictive chaotic model leading to an increase in accuracy by 67%; data assimilation using NARX network gives 553% improvement; and multi-model ensemble predictions using FTDNN with batch learning is the most effective method to improve the performance of predictive chaotic model by 967%. Nevertheless, additional case studies might be needed to test further the reliability of the improved methods and the possibilities of combining them.

Overall, the presented research makes a contribution to developing more accurate methods of surge prediction. The modeling techniques based on the methods of nonlinear dynamics, chaos theory, statistics and neural networks with several enhancements and innovations have demonstrated that the predictive chaotic model can serve as an efficient tool for accurate and reliable short-term predictions of storm surges in order to support decision-makers for flood prediction and ship navigation. We believe this approach has a very good potential to become a complementary method used by practitioners along with the traditional numerical ocean models.

Delft, 6 December, 2011

Michael Siek

Table of Contents

List of Figures

List of Tables

Acknowledgements

Over some years, it has been my fortune to meet many nice people who have provided me with their time, knowledge, support and companionship and patience. This thesis is a complex result of a long process involving chaotic brainstorms, discussion, feedbacks and motivations from these people.

First of all, I would like to thank to my promotor and also my supervisor, Professor Dimitri Solomatine, for providing me to do this research. He has not only given me with brilliant ideas, deep thought, fresh innovation, critical feedbacks, but he also offered me full support and strong motivation. His sense of humour and nice communication made doing research with him attractive and pleasant. Without these contributions from him, I would never have come to this stage.

I would also thank to Professor Roland Price who was my promoter in the first year. He has given me a number of bright ideas and understanding on coastal modeling techniques and provided some inputs and suggestions for my research.

One important person with a great contribution to this research is Dr. Slavco Velickov who did his Ph.D. research in the field of nonlinear dynamics and chaos theory. Slavco, thank you for your patience and initial help me in chaotic modeling.

I would also like to thank Dr. Martin Verlaan and Regien Brouwer from Rijkswaterstaat and Deltares for their patience to explain how the Dutch storm surge model (DCSM/Waqua) and other European storm surge models work, and for their kindness in providing access to Matroos data sets. They also have brought new views and ideas in the field of storm surge prediction not only in theory but also in real operational model and practical perspectives.

My thanks also go to Oscar Hernandes, a former M.Sc. student in Hydroinformatics whom I co-supervised his M.Sc. research. He provided a contribution to the optimization of a chaotic model.

I am also very grateful to Jan Luijendijk, Head of Hydroinformatics and Knowledge Management Department for his continuous supports in all aspects and creating pleasant and wonderful working environment.

I am thanking also Professor Arthur Mynett for his warm smile and nice chat. Thanks to Biswa Bhattacharya for being a nice roommate in the office and for the useful discussion, Arnold Lobbrecht, Andreja Jonoski, Ioana Popescu, Schalk Jan van Andel, Jos Bult and all other lecturers and staff of UNESCO-IHE for all ideas, conversations and support in socio-technical matters.

My great gratitude goes to all members of Delft Cluster Project and partners which have funded this research.

I would like to thank Ph.D. and M.Sc. colleagues, Durga Lal Shrestha, Adrian Almoradie, Nagendra Kayastha, Carlos Velez and all of my friends for their help, pleasant conversations and creating a cheerful atmosphere.

I am deeply indebted to my grandma, mother, brother and sister for their continuous support and enthusiasm. I would thank to my uncle and auntie who live in the village of Lunteren for kindly bringing me all what I needed here. Last but not least, my gratitude and great appreciation set out to Liliani for her deep heart and steady attention and to my cousin Yvonne Yu for her support and allowing me to play her luxurious violin during hectic moments.

Michael Siek

CHAPTER 1: INTRODUCTION

"The scientist does not study nature because it is useful to do so. He studies it because he takes pleasure in it, and he takes pleasure in it because it is beautiful. If nature were not beautiful it would not be worth knowing, and life would not be worth living."

H. Poincaré

This chapter introduces the main problems, motivations and objectives of the research. Storm surges have become one of the most disastrous natural events. The mechanism of storm surge generation and possible model predictions using hydroinformatics tools, such as physically-based and data-driven modeling are discussed. Previous and recent developments of chaotic modeling in predicting aquatic phenomena are explored. Several improvements on building a chaotic model from time series, advances in solving some issues, data assimilation and multi-model ensembles are proposed.

1.1   Motivation: Natural disasters

Natural disasters have been occurring since the earth or universe was created by a big bang. Humans are mostly concerned with earthquakes, flooding, volcanic eruption and landslides. Naturally, a disaster has profound environmental effect, human loss and frequently incurs financial loss. One of most disastrous events is storm surge that can cause floods in the coastal areas. Storm surge is a meteorologically forced long wave motion which is pushed toward the shore. It is generated by a combination of meteorological forces of the wind friction and low air pressure due to a storm and oscillates in the period range of a few minutes to a few days (Gonnert *et al.*, 2001). In the ocean, local wind waves can add to the water level, and the storm surge can be amplified (or reduced) by interference with the strictly regular astronomical tides. Extreme coastal floods can be related to extreme storms, like cyclones or hurricanes which attack the open coast. In some coastal areas, such floods can be generated by unusual sequences of wind set-up and air pressure variations. In addition, wind driven waves can be superimposed on the storm tide. This rise in sea level can cause severe flooding in coastal areas, particularly when the storm tide coincides with the high tides (Battjes & Gerritsen, 2002).

Figure 1-1: The 1953 North Sea floods due to a heavy storm resulting in severe destruction in the coastal areas (source: deltawerken).

Over the past centuries, a number of severe coastal floods have destructed many places in the world. For example, the 1953 North Sea flood is a major flood caused by a storm tide, a combination of a high spring tide and a severe European windstorm (Figure 1-1). In combination with a tidal surge of the North Sea the water level locally exceeded 5.6 meters above mean sea level. The flood and waves overwhelmed sea defenses and caused extensive flooding. The flood struck the Netherlands, Belgium, England and Scotland. Large part of Dutch area is located below mean sea level and relies heavily on sea defenses, effecting on 1,835 deaths. Most of these casualties occurred in the southern province of Zeeland. In England, 307 people were killed whereas 28 people were killed in West Flanders, Belgium. Further loss of life exceeding 230 occurred on watercraft along Northern European coasts as well as in deeper waters of the North Sea; the ferry MV Princess Victoria was lost at sea in the North Channel east of Belfast with 133 fatalities, and many fishing trawlers sank.

Figure 1-2: Maeslant and Oosterschelde storm surge barriers (source: deltawerken).

In the Netherlands, after the storm surge flood of 1953, some actions were undertaken to increase the safety of the delta areas in the long run. Although most delta areas need to be closed for safety, several seaways should stay open because of the economic importance of the Rotterdam and Antwerp harbours. Dikes and storm surge barriers in the delta areas and along the Dutch coast have been constructed or systematically improved with a return period of 1 in 10,000 years (Figure 1-2). The existence of these mega structures induces a challenge of how to operate these properly since they have to be open or closed on the right time to avoid the barrier breaking. An accurate and reliable predictive storm surge model is critically required for this purpose.

1.2 Modeling Natural Phenomena: Hydroinformatics

A model is defined as a simplified representation of real world with an objective of its explanation or prediction. Modeling includes studying the system, posing the problem, collecting data, preparing and building the model, testing it, using it, interpreting the results and possibly reiterating (Solomatine, 2002).

One of the fundamental modeling technologies of modern science that can enhance our understanding of complex natural phenomena and its processes is mathematical modeling (often with numerical simulations). This kind of modeling technique is often called physically-based modeling. The construction of this model is primarily based on the conceptualizations of physical processes and behaviors of a particular natural phenomenon that are expressed in mathematical equations typically implemented as algorithms. Such equations describe the quantitative processes of the whole system based on fundamental principles, such as conservation of mass, momentum and energy. The solution of these equations requires the application of specific numerical techniques and the imposition of certain boundary condition analysis. This branch of science that focuses on the discretisation of the physical domain and the corresponding equations governing the natural processes is known as computational hydraulics. A number of contributions on identifying and solving water and environmental problems can be found, for instance: oceanic waves, flood forecasting, sediment transport and morphodynamics. A mathematical model based Navier-Stokes equation is often used for these applications. An accurate quantitative description of the causal relationships between processes, actions and consequences in the water systems can be obtained from studies of mathematical models.

The developments of computational hydraulics techniques have reached to the uses of information and communication technologies (ICT) which comes from the other discipline of information or computer science. The merging of these different disciplines and technologies has lead to emergence of a new field of hydroinformatics (Abbott, 1991). A hydroinformatics system is an electronic knowledge encapsulator that models part of the real world and can be used for the simulation and analysis of physical, chemical and biological processes in water, for a better management of the aquatic environment. Hence, the development of mathematical models, which adequately represent our current image of reality is at the heart of hydroinformatics (Price, 2001).

In the applications, hydroinformatics draws on and integrates hydraulics, hydrology, environmental engineering and many other disciplines. It considers application at all points in the water cycle from atmosphere to ocean, and in artificial interventions in that cycle such as urban drainage and water supply systems. It provides support for decision making at all levels from governance and policy through management to operations. This may involve many individual components and processes, interacting with each other in complex ways. Thus, it requires a successful collaboration of experts from many different disciplines. The result can lead to the convergence of different sciences and the notion of integrated modeling.

In addition to computational hydraulics, hydroinformatics has a strong interest in the uses of other techniques based on the analysis of the data characterizing the underlying system. The model based on these approaches is primarily defined on the basis of connections between system state variables (input, internal and output variables) with only a limited knowledge of the details about the physical behavior of the system. Such models can be called data-driven models. During the recent decade, such models became quite popular due to the redundant availability of data. The approaches in data-driven modeling generally originate in statistical methods and artificial intelligence. Several popular techniques include: artificial neural network (ANN), instance based learning, model tree, Bayesian learning, committee machine, fuzzy rule based system and genetic programming. These techniques are used to build data-driven models based on the analysis of all the data characterizing the system under study.

By the rapid developments and recent research, nowadays the field of hydroinformatics has been broaden into emerging socio-economic aspects. The inherently social-economic nature of the problems in water management and decision making processes and how to bring technologies into use are essential matters. This leads to not only about technologies but also the modeling of socio-economic issues, for example modeling socio-economic consequences of certain activities in the aquatic environment which involve different stakeholders and public participation in the decision making processes in the water resources management. Since the problems of water management are severe in most countries, while the resources to obtain and develop solutions are minor, the necessity to examine these social-economic processes is crucial.

Besides the issues in socio-economic field, there are specific technical challenges and limitations on the use of computational models for describing natural dynamical systems in which many elements are interacting each other in complex manners. For example, the derivation of the hydrodynamic equations has to make certain assumptions due to limited knowledge of the underlying processes, such as the bed resistance. Such assumptions are usually expressed in empirical forms that require the values of one or more parameters to be identified through calibration process. The model results of a computational model have to agree closely with the observed data. The physical integrity of the parameters should not be violated during calibration process. The mathematical model is an approximate representation of the real world system. The larger errors of model estimation or predictions are mostly expected as prediction horizon increases. Several sources of model errors include: missing processes and parameters in the model, governing laws of the physical processes due to limited knowledge, the error due to assumptions in equation discretization, errors in the measured data, imprecise estimation of initial and boundary conditions, computer round-up error and so forth. Therefore, uncertainty analysis is required to provide the confidence level of computational model predictions or results.

1.3   Predicting Storm Surges

The mechanism leading potentially to coastal floods is well understood, given the configuration of the coastline and the bathymetry, the severity of the storm surge depends primarily on wind speed, wind direction and duration. The meteorological conditions are affected by the path and the velocity of the depression systems, moving across the Sea. When winds push water towards the coast, it tends to accumulate into what is commonly referred to as storm surge. If a particular high surge occurs together with a tidal maximum, both effects accumulate and serious flooding can result, depending on the coastal structure and their protection.

The analyzes of the risk of coastal floods due to storm surges are not straightforward, because an observed flood is not a single independent event in statistical terms. Rather, the flood is a consequence of a set of different determinants, like tides, wind and air pressure, or of a set of sequences of these factors. Both the mean sea level and the flood height will vary along the coast and the risk of coastal flood depends on emergency preparedness planning and the design of coastal facilities and structures, such as flood embankments. The ocean water level variations due to various determinants and their complex interactions show long-term persistence leading to the correlated extreme events (Alexandersson et al., 1998; Butler et al., 2007).

1.3.1 Physically-based modeling

In general, astronomical tides have the large contribution to the ocean water level variations in open oceans and many well-exposed coasts. Traditionally, the analysis of water levels usually employs linear methods that decompose sea levels into tides and other (usually meteorological) components. The amplitudes and phases of the tidal constituents driven by the astronomical motion of the Earth, Moon and Sun (with known periods) can be estimated by using Fourier analysis, response analysis or linear regression methods. In particular, the weakly nonlinear shallow water waves like storm surge can be represented by the Korteweg-de Vries (KdV) equation (Korteweg & de Vries, 1895) which is an exact solvable partial differential equation. The KdV equation can be obtained in the continuum limit of the Fermi-Pasta-Ulam Experiment (Fermi et al., 1955). The solitary wave solutions have behavior similar to the superposition principle, despite the fact that the waves themselves are highly nonlinear (Zabusky & Kruskal, 1965). In real applications, however, the water level dynamics in coastal and estuarial swallow-water areas, such as the Dutch coast, may differ significantly from the astronomical estimated constituents (superposition principle) – due to the nonlinear effects that include meteorological forcing, tidal current interactions, tidal deformations due to the complex topography and river discharges (Prandle et al., 1978).

Essentially, the coastal floods due to storm surges can be predicted with an accuracy that depends on the accuracy of the meteorological forecasts. An appropriate numerical weather model can predict the motion of atmospheric depression with a satisfactory accuracy in a range of several days. The wind and air surface pressure fields predicted by this model can be utilized as some driving forces of the sea motion in a shallow water model allowing for storm surge predictions. In the Netherlands, the storm surge predictions are made by a shallow water model so-called Dutch Continental Shelf Model (DCSM) which receives meteorological prediction as inputs from a numerical weather prediction (NWP) system so-call High Resolution Limited Area Model (HIRLAM) (de Vries, 1991; Gerritsen et al., 1995; Unden et al., 2002). The storm surge predictions from DCSM and astronomical tidal predictions made by means of harmonic analysis are then added up to attain total sea water level predictions (see Figure 1-3).

Over the past two decades, the operational numerical storm surge models have significantly been improved, which turns out to be very essential to anticipate the occurrence of coastal flooding. A number of advances on physically-based storm surge modeling have been reported by (Bode & Hardy, 1997; Heemink et al., 1997; Battjes & Gerritsen, 2002; Verlaan et al., 2005). These improvements include: refining computational grids, utilizing more accurate calibration of models with better data, using an improved numerical schemes and incorporating data assimilation technique into the model.

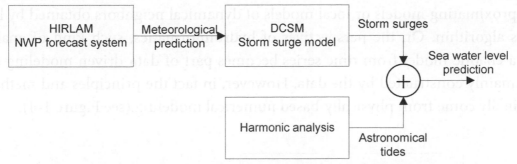

Figure 1-3: Sea water level forecasting system in the Netherlands.

1.3.2 Data driven modeling: Nonlinear dynamics and chaos theory

The ideas of dynamical chaos were firstly introduced by Poincaré when he entered a contest in 1887 and tried to solve a generalization of the famous three body problem, which was considered one of the most difficult problems in mathematical physics. His incomplete solution leads to a new era on chaos in the solar system. Some decades after, around 1960, Edward Lorenz made a discovery while working in his laboratory at MIT on studying the full equations describing atmospheric flow and weather phenomena. This experiment led him to conclude that a small perturbation of the initial conditions can lead to enormous differences over time. He coined the concepts of chaos theory as a new field in dynamical system. Several properties of chaos include: determinism, small number of variables, complex behavior, low dimensional in phase space, bifurcations, strange attractor and sensitivity to initial conditions. Some nonlinear differential equations can exhibit chaotic properties.

After the discovery of chaos, the developments in the methods of nonlinear dynamics and chaos theory were progressing fast with numerous findings on chaotic behaviors in many dynamical systems which previously were believed to be random behaviors. Several nonlinear methods, such as: correlation dimension (Grassberger & Procaccia, 1983a), Cao's method (Cao, 1997), Lyapunov spectrum (Sano & Sawada, 1985), method of time delays (Takens, 1981), mutual information (Fraser & Swinney, 1986) have been introduced for identifying the existence of chaos and the properties of the dynamical systems which are mostly derived from differential equations. Such developments have provided a set of nonlinear chaos analysis tools. Using the same principles and methods, the nonlinear analysis can be used for a time series from real observations of natural phenomena (i.e. rainfall, weather) instead of from a set of differential equations. Recent developments allow for building a chaotic model from observed time series and making predictions in case of the presence of deterministic chaos in the dynamical system. The chaotic model from time series can be built by embedding the original time series in high-dimensional time-delayed phase space. The proper values of time delay and embedding dimension can be determined by means of the nonlinear chaos analysis. Predictions of a chaotic model are made by either global approximating models or local models of dynamical neighbors obtained by k-nearest neighbors algorithm. On the perspectives of hydroinformatics, such kind of analysis and building a chaotic model from time series becomes part of data-driven modeling since the model is mainly constructed by the data. However, in fact the principles and methods used here originally come from physically-based numerical modeling (see Figure 1-4).

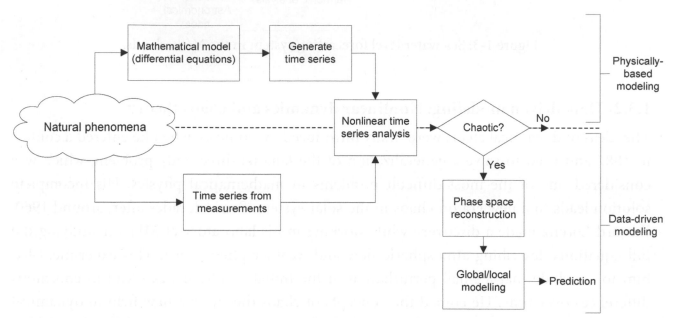

Figure 1-4: Different processes of chaotic model development between physically-based modeling and data-driven modeling.

1.3.3 Main relations between the two modeling paradigms: chaotic modeling

Complexity of the described phenomena prompts for adequate methods to describe them, and one of them is chaos theory. Primary links between the methods of nonlinear dynamics and chaos theory and the numerical storm surge model can be described as follows. The basis of the widely used physically-based numerical storm surge model is the Navier-Stokes shallow water equations, stating the physical laws of mass and momentum conservations (Dronkers, 1964). These equations are inherently nonlinear. The sensitive dependence on the initial and boundary conditions of the dynamical evolution of such systems, and the broadband and continuous power spectra are the indicators of deterministic chaos. A proof on the existence of chaotic behavior in Navier-Stokes equations and turbulence has been presented by Li (2007). The presence of bifurcations in ocean, atmospheric and climate models for understanding the variability of oceanic and atmospheric flows as well as the climate system has been investigated and analyzed by Simonnet *et al.* (2009). As models, chaos dynamical systems show rich and even surprising variety of dynamical structures and solutions. Most appealing for researchers and practitioners is the fact that the deterministic chaos provides a prominent explanation for irregular behavior and instabilities in dynamical systems (including storm surges), which are deterministic in nature.

The most direct link between the concept of deterministic chaos and the data-driven modeling is the analysis of data (time series) from observations using the well-developed methods of nonlinear dynamics and chaos theory. Important contributions in this area were made by (Tsonis, 1992; Abarbanel, 1996; Kantz & Schreiber, 2004; Donner & Barbosa, 2008). Note that this approach is, in fact, data-driven, since it is based on the analysis of the observation data rather than the explicit mathematical analysis of the properties of the underlying equations.

1.4 Chaotic Behaviors in Ocean Surge and Other Aquatic Phenomena

The presence of chaotic behaviors in aquatic phenomena (i.e. ocean, weather, rainfall) have been studied and reported for more than a century. Chaos in water motions and waves was initiated by (Korteweg & de Vries, 1895) through modeling weakly nonlinear shallow water waves using an exact solvable partial differential equation. Edward Lorenz investigated the atmospheric model and found the principles of chaos theory and sensitivity to initial conditions (Lorenz, 1963). Floris Takens and David Ruelle investigated the fluid turbulence and introduced how to reconstruct a dynamical system from observed time series (Takens, 1981). They coined the famous term "strange attractor".

In the field of hydrology, the presence of chaotic behavior in rainfall has become main research interest in last two decades. A number of methods are available to identify the existence of chaos in hydrological time series. The possible chaotic behavior in the temporal rainfall of storm events and the limit of its predictability were studied by Rodriguez-Iturbe *et al.* (1989). Jayawardena & Lai (1994) explored the methods of chaos theory and nonlinear and linear (ARMA) prediction methods for streamflow and rainfall data series in Hong Kong. Storm term prediction of rainfall, estimation correlation dimension with sufficient data and the use of inverse method have been investigated by Sivakumar *et al.* (1999) for the case study in Singapore. Correlation dimension method has been the most widely used in rainfall and other hydrologic time series, either as a method of proof or as a method of preliminary evidence. However, due to the potential limitations of this method, criticisms on its application to hydrologic time series and the reported results have been reported by Ghilardi & Rosso (1990) and Schertzer *et al.* (2002). In addition to the problems of insufficient data size, sampling frequency and presence of noise, the analysis of rainfall (and other hydrologic) time series, finer-resolutions in particular, and the outcomes might be significantly influenced.

In the area of oceanography, the early nonlinear analyzes of the ocean water levels at the Florida coast have been conducted by Frison *et al.* (1999); this work was for us an important motivation. The use of chaotic model for predicting the errors of a deterministic numerical model for ocean water level at Venice Lagoon has been reported by Babovic *et al.* (2000). Several examples of using predictive chaotic model (CM) for storm surge predictions were reported by Solomatine *et al.* (2000) and Walton (2005) using univariate local models. The methods for building multivariate predictive chaotic models was extended in the PhD study by Velickov (2004) and the results showed that they can provide accurate short-term predictions. In addition, the applications of artificial neural networks to oceanography (i.e. predicting ocean waves) have also been explored by Deo (2010) and Zamani *et al.* (2008). Modeling of the impacts of storm surges to dunes has been investigated by Roelvink *et al.* (2009). Figure 1-5 illustrates the research developments by several authors on the uses of methods of nonlinear dynamics and chaos theory for aquatic applications.

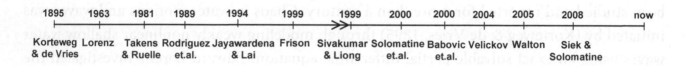

Figure 1-5: Some of the research developments on chaotic modeling of aquatic phenomena.

1.5 Main Objectives

Previous studies and research on nonlinear dynamics and chaos theory and data-driven modeling for water-related applications have shown very promising results and this can also be open research for many other fields of applications, for instance: transportation, robotics, biomedical and electronics applications. However, several issues are unsolved or need for further research. The existing methods have a number of advantages and disadvantages and these provide some space for improvements. This research is a combination of earlier works with more data and have several improvements on building predictive chaotic model from observed time series.

The main research objective related to the Ph.D. research project are to build more accurate chaotic (data-driven) model that can serve as a complementary model to the existing operational storm surge models, with the North Sea region being the main case study. Some experiments optimization of chaotic models were also conducted using the data on surge at San Juan tidal station (Puerto Rico) in the Carribbean Sea. The main objective can be further detailed as follows (see Figure 1-6):

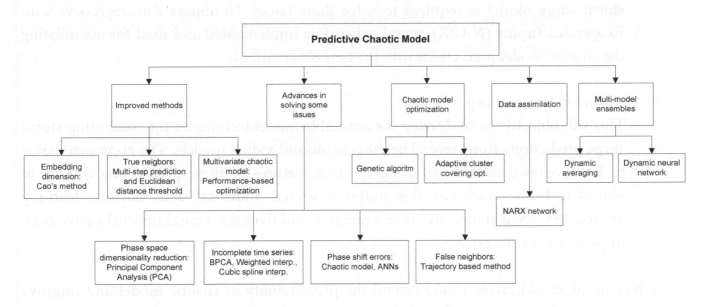

Figure 1-6: Schematic diagram of several issues, methodologies and main objectives of the research.

1. Improved techniques on building predictive chaotic model

   A number of open issues in standard methods should be solved in order to improve the accuracy of predictive chaotic models, such as: determining embedding dimension and time delay, non-optimal selection of chaotic model parameters, false neighbors, data with missing values, high dimensionality of phase space, phase prediction error. For instance, the approaches using Cao's method and genetic algorithm and adaptive cluster covering optimization (ACCO) techniques are utilized to estimate the values of embedding dimension and time delay as well as number of neighbors. Removing false neighbors can be made by introducing multi-step prediction and trajectory based method. Several methods, like: weighted sum of linear interpolations, Bayesian principal component analysis (BPCA) and cubic spline interpolation are implemented for building predictive chaotic model from incomplete time series. Dimensionality reduction using principal component analysis (PCA) is proposed to bring the information (distance and neighbors) in high dimensional phase space into lower one and able to remove data noises. Performance-based optimization is used for choosing optimal variables and parameters in multivariate chaotic model. Models, such as: ANN and chaotic model, are utilized for correcting the phase errors. The performance of model results are compared with global ANN model performance.

2. Incorporating data assimilation methods

   The unsolved issues on imprecise determination of initial condition and round-up computer errors results in decreasing prediction performance after certain time of prediction horizon. Developing a real-time data assimilation technique for chaotic storm surge model is required to solve these issues. Nonlinear Autoregressive with Exogenous Inputs (NARX) neural network is implemented and used for assimilating the chaotic model predictions with the new observations.

3. Multi-model ensemble predictions

This sub objective is to develop the ensemble model techniques for combining storm surge predictions from several heterogeneous individual models. The approach that is to be developed focuses on using the best features of these predictions coming at almost-real-time rate. For this purpose, several statistical and machine learning approaches, for instance, dynamic averaging and dynamic neural networks have been implemented and tested.

Achieving these objectives would extend the predictability of chaotic model and improve the accuracy for short and medium-term predictions. Finally, the model can complement to the existing operational storm surge model in supporting decision makers for the purposes of flood forecasting and ship navigation.

One other promising improvement on storm surge prediction could be achieved if the meteorological predictions coming from NWP model (i.e. HIRLAM) instead of using only meteorological observations are included as the inputs of predictive chaotic model (Figure 1-3 and Figure 1-4). This technique can produce a hybrid model and extend the predictability of chaotic model. However, this idea could not have been further explored at this stage due to lack of time and unavailability of the high frequency meteorological prediction data.

1.6 Thesis Outline

This thesis is arranged along the following structure:

Chapter 1 introduces the problems, motivations, objectives of the research with a brief description of the previous research and possible improvements and solutions of the existing methods.

Chapter 2 describes the case study in North Sea covering the ocean dynamics, tides and sea level, storm surge condition, warning service, procedure issuing alarms and data description used for this research.

Chapter 3 presents the basic principles of physical oceanography and storm surge modeling, the European operational storm surge models, North West Shelf Operational

Oceanographic System, ECMWF and the ways of linking predictive chaotic model with the existing European operational storm surge models.

Chapter 4 describes the techniques of computational intelligence as the major tools for data-driven modeling. Several topics, like: artificial neural network (ANN), instance-based learning, hierarchical modular model and evolutionary and other randomized search algorithms.

Chapter 5 explains the main methods of nonlinear dynamics and chaos theory which are used for building storm surge model. It also describes the discovery of chaos, basic of chaos and its properties, chaos in iterative maps, chaos in differential equation, phase space reconstruction, finding proper time delay and embedding dimension, Lyapunov exponents, chaotic model prediction and recurrence plots.

Chapter 6 describes how to find the proper values of delay time and embedding dimension, how to build a predictive chaotic model for storm surges and make predictions. A number of nonlinear time series analysis are explored, like: power spectrum, Poincaré section, mutual information, correlation dimension, false nearest neighbors, Cao's method, Lyapunov spectrum and Kaplan-Yorke dimension. Global and local modeling, direct and multi-step prediction and the use of CI techniques for building predictive local and global models are discussed here.

Chapter 7 explores several enhancements on building predictive chaotic model, include: phase space dimensionality reduction with principal component analysis (PCA); prediction error correction with predictive chaotic model (PCM) and artificial neural network (ANN); building predictive chaotic model from incomplete time series with weighted sum of linear interpolation, Bayesian PCA, cubic spline interpolation; and finding true neighbors using trajectory based method.

Chapter 8 presents the optimization of predictive chaotic model parameters: delay time, embedding dimension and number of neighbors. Three optimization methods are utilized, include: grid search (GS), genetic algorithm (GA) and adaptive cluster covering optimization (ACCO).

Chapter 9 describes a real-time data assimilation technique for chaotic storm surge model using NARX neural network. The modeling results are compared with several European operational storm surge models.

Chapter 10 presents the chaotic multi-model ensemble prediction methods in high
      dimensional space, include: dynamic averaging, MLP-ANN and FTDNN with batch
      and incremental learning.

Chapter 11 summarizes the conclusions and recommendations.

CHAPTER 2: CASE STUDY

"Nordsee ist Mordsee"
Film title, 1976

The case study of this research concentrates on predicting the storm surges in the North Sea, specifically at Hoek van Holland as the entrance of Rotterdam Harbor. This research is executed by collaboration between UNESCO-IHE and Rijkswaterstaat/Deltares and partly funded by Delft Cluster Project on "Safety against floods". Rijkswaterstaat/Deltares with NOOS data exchange facility provides observation and prediction data, such as sea water level and surge along the coastlines of North Sea. This involves several meteorological institutions from European countries surrounding the North Sea, include: Netherlands, Germany, UK, Denmark, Norway and Belgium.

2.1 Study Area: The North Sea

The North Sea lies between Norway, Denmark, Germany, the Netherlands, Belgium, France and Great Britain. It links up with the Atlantic Ocean to the north and also the southwest, via the Channel. To the east it links up with the Baltic Sea. The total surface area is approximately 750,000 km² and the total volume 94,000 km³. The North Sea has a dynamically active regime dominated by strong tides and frequent passages of mid-latitude synoptic weather systems (Droppert, 2001). The waters are mostly shallow (depth < 150 m) in the region.

In the Netherlands accurate prediction of storm surges is very importance against the possible coastal flooding since large areas (about 55%) of the land lie below sea level. These below sea level areas are most densely populated and important economy areas. Since the disastrous storm of 1953, the dikes and dams in the Delta area and along the rest of the coast have been systematically improved. The dikes and barriers are designed to withstand very severe storm surges that they will occur on average only once in every 4,000 to 10,000 years. Less extreme weather can also damage the dikes through the impact of breaking waves and strong tidal currents (Corkan & Council, 1948; Carter, 1985).

The reliability of North Sea coastal defences is checked regularly and systematically by the National Department of Public Works (Rijkswaterstaat), the provincial public works authorities, the water boards and the municipalities. Supreme control is in the hands of Rijkswaterstaat, which is also responsible for high tide warnings. The provincial governors and the authorities in charge of the dikes and dams must be warned when the tidal rise is expected to reach dangerous levels due to a combination of tidal movements, high river discharges and winds (Verlaan *et al.*, 2005).

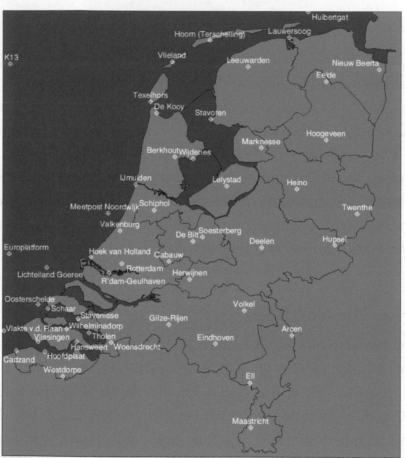

Figure 2-1 North Sea region and the position of the important meteorological stations.

In the North Sea, high tides occur approximately every twelve hours. The main tidal stream enters the North Sea along the Scottish coast; a less important stream comes in through the Channel. As a result, the difference in level between high tide and low tide is not the same everywhere. Each point on the coast has its own tidal difference, i.e. 3.80m on average in Vlissingen. The actual tidal difference depends not only on the positions of the sun and the moon. It is also determined by the weather, and primarily by the wind. North- westerly storms are notorious in this respect. They can blow the full length of the North Sea without interruption and sweep the water up against the Dutch coast. The rise on any particular occasion depends on the direction, the force and the duration of the storm (Langenberg *et al.*, 1999; Howarth *et al.*, 2001).

There are seven important locations considered in the research include: Delfzijl, Euro platform, Haringvliet 10, Hoek van Holland, K13 platform, Vlissingen and Ijmuiden. The variables (observations) used in modeling are: astronomical water level, surge water level, wind speed & direction and air pressure. The observations are available with sampling time 10 minutes and 1 hour from January 1990 till March 1996. However, the observation will be extended till December, 2006 in order to examine the resulting models in predicting extreme storm surges in November, 2006.

2.2 North Sea Characteristics

2.2.1 Ocean dynamics

As tides from the deep Atlantic Ocean enter the North West European shelf, they propagate around the coast in the form of long gravity waves. Conservation of energy flux requires an increase in tidal height and current amplitude as water depths decrease. The increase in tidal currents gives rise to strong bottom friction and generation of intense turbulence, dissipating a large amount of energy. It has been estimated that the North West European shelf accounts for about 10% of the global shallow water tidal dissipation (Droppert, 2001).

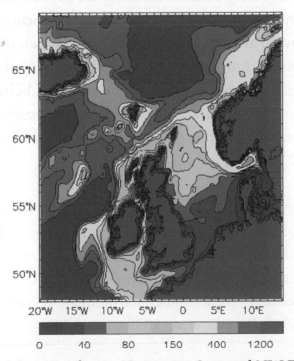

Figure 2-2: Bathymetry of the NE Atlantic, Norwegian Sean and NW European Shelf (Droppert, 2001).

The North Sea has moderate fetch for easterly or westerly wind directions, and a long fetch for northerly winds. The highest recorded waves have been generated by northerly winds, for example significant wave heights up to 11m in the central North Sea in early January 1995. Waves of return period 50 years have significant wave height 16 m in the Northern North Sea and 8 m in the South. Within the North Sea and North West European shelf waters, bottom friction is important in limiting growth of the longer period waves (e.g. waves of around 7 seconds over Dogger Bank), and this must be accounted for in numerical wave models (Carter, 1985; Bijl, 1997).

2.2.2 Tides and sea level

As tides from the deep Atlantic Ocean enter the NW European shelf, they propagate around the coast in the form of long gravity waves. Conservation of energy flux requires an increase in tidal height and current amplitude as water depths decrease. The increase in tidal currents gives rise to strong bottom friction and generation of intense turbulence, dissipating a large amount of energy and mixing the water column. It has been estimated that the NW European shelf accounts for about 10% of the global shallow water tidal dissipation (Carter, 1985; Droppert, 2001).

The combined effects of Coriolis and frictional forces and the geometry of the NW European shelf result in complex tidal patterns in this region. In the semi-enclosed North Sea, for example, the tide originating from the North Atlantic enters from the north as a progressive Kelvin wave, travelling southward along the eastern side of the UK coast. Much of the tidal energy is dissipated in the Southern Bight, but a portion is reflected as a damped wave, propagating northward along the continental coast. When the incoming and reflected Kelvin waves are superimposed together, three amphidromic systems are established in the North Sea. The one in the Southern Bight lies about halfway between East Anglia and the Netherlands. The two further North are displaced progressively eastward from the mid-distance as the reflected wave is damped gradually when travelling northward (Debernard *et al.*, 2002; Butler *et al.*, 2007).

The actual observed tides are in fact more complex than this, when tidal constituents other than M2 are considered. Superposition of semidiurnal M2 and S2 tides for example gives rise to a spring/neap cycle that has a period of about 14 days. In addition, when water becomes shallower, higher harmonics of astronomical constituents are generated by bottom friction and non-linear effects. These higher harmonics are referred to as shallow water constituents (Dronkers, 1964; Prandle *et al.*, 1978).

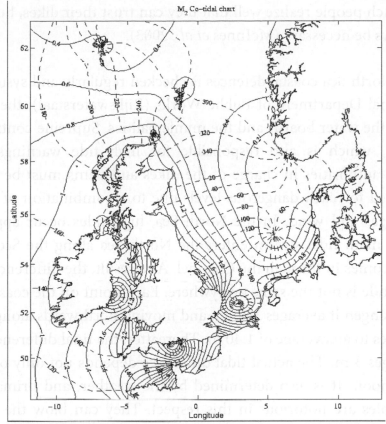

Figure 2-3: M2 co-tidal plot for North West European shelf seas (Droppert, 2001).

The tidal range on the continental coast of the North Sea is much smaller, rendering the impact of a storm surge that much greater. Though changes in sea-level are predominately due to tides in the NW European shelf, winds and variations in the atmospheric pressure can raise or lower sea level by up to several meters, producing a storm surge. A wind-driven current can cause a rise in the sea level by piling up water against the coast. Changes in atmospheric pressure on the other hand give rise to an 'inverted barometer' effect: a fall in pressure by 1 mb resulting in a 1 cm rise in sea-level. Therefore, if a positive peak surge (higher water level) occurs at the time of high tide, flooding may result along the coastal areas. The much-publicized North Sea storm surge in 1953 occurred when the high spring tide interacted with a deep depression which first travelled eastward to the north of British Isles and then south-eastward into the North Sea. In this surge event, the east coast of British Isles and the Netherlands were severely flooded, and about 2000 people lost their lives (Bijl, 1997; Debernard *et al.*, 2002; Beniston *et al.*, 2007).

2.3 Storm Surge Condition in the North Sea

Since the great flood disaster of 1953, when a combination of a spring tide and storm force north-westerly winds led to the emersion of large parts of Zeeland and South Holland, the Dutch sea defences have always proved reliable. Even so, extreme weather conditions still pose a danger. Dutch people realize well that they can trust their dikes, but not blindly, and vigilance will always be necessary (McInnes *et al.*, 2003).

The reliability of North Sea coastal defences is checked regularly and systematically. This is done by the national Department of Public Works (Rijkswaterstaat), the provincial public works authorities, the water boards and the municipalities. Supreme control is in the hands of Rijkswaterstaat, which is also responsible for high tide warnings. The provincial governors and the authorities in charge of the dikes and dams must be warned when the tidal rise is expected to reach dangerous levels due to a combination of tidal movements, high river discharges and wind. In the North Sea, high tides occur approximately every twelve hours. The main tidal stream enters the North Sea along the Scottish coast; a less important stream comes in through the Channel. As a result, the difference in level between high tide and low tide is not the same everywhere. Each point on the coast has its own tidal difference. In Vlissingen it averages 3.80 m, and moving northwards along the coast to Den Helder it diminishes to an average of 1.40 m. Thereafter, the tidal difference increases again: at Delfzijl it averages 3 m. The actual tidal difference depends not only on the positions of the sun and the moon. It is also determined by the weather, and primarily by the wind. North- westerly gales are notorious in this respect. They can blow the full length of the North Sea without interruption and sweep the water up against the Dutch coast. The rise on any particular occasion depends on the direction, the force and the duration of the gale (Verlaan *et al.*, 2005).

Since the disastrous storm of 1953, the dikes and dams in the Delta area and along the rest of the coast have been systematically improved. The recent completion of the storm surge barrier in the New Waterway was the last piece in the jigsaw of protective measures. Apart from the Western Scheldt, the entire delta can now be sealed off from the sea by huge storm surge barriers. The dikes and barriers are designed to withstand 'very severe storm surges': that is to say, storm conditions so rare that they will occur on average only once in every 4,000 to 10,000 years. Even though such super storms are highly unlikely in our own lifetime, we cannot afford to relax our guard. Less extreme weather can also damage the dikes through the impact of breaking waves and strong tidal currents.

Whenever dangerous high tide levels are anticipated, it is the duty of the Storm Surge Warning Service (SVSD) to notify the dike and dam authorities and other bodies responsible for public safety. The SVSD is always ready to spring into action, 24 hours a day. During gales, it keeps meticulous watch on developments in coastal tide conditions, particularly if the wind direction is between south- westerly and northerly. It also predictions critical high tide levels and issues advance warnings to the relevant authorities. The SVSD itself takes no measures to defend the dikes - that is up to the local dike boards.

There are three rise levels that prompt the SVSD to come into action: pre-warning, warning, and alarm. Because the timing of high tides varies from one place to another and because a gale will seldom affect the whole coastline with equal force, the coastal region is divided into sectors. In each sector there is a reference station (Battjes & Gerritsen, 2002).

2.3.1 Storm Surge Warning Service

The Storm Surge Warning Service (in Dutch abbreviated as SVSD) is responsible for the alert of the dike and dam authorities and other relevant bodies in the Dutch tidal region whenever a storm surge is expected. This will allow them to take appropriate measures. The SVSD is a service provided jointly by the Department of Public Works (Rijkswaterstaat) and the Royal Netherlands Meteorological Institute (KNMI). The management of the SVSD is in the hands of Rijkswaterstaat's National Institute for Coastal and Marine Management (RIKZ). Weather and the use of weather products like wind, pressure, temperature and model-results like wind- and pressure fields are very important for the Storm Surge Warning Service. The digital products are input for the hydro-dynamic models. With these models, the water level predictions are made. The graphical weather products are used for interpreting the model outcome. For the interpreting predictions and measurements have to be combined. Products on different prediction periods are used. A recent development has been the ensemble prediction for wind surge, where the 50 scenarios of the ECMWF ensemble prediction are used as input for a wind surge model. The probabilistic outcome of this model has proved very useful for planning purposes (Heemink *et al.*, 1997; Verlaan *et al.*, 2005).

2.3.2 Procedure for issuing warnings and alarms

Every day the Hydro Meteorological Centre in Hook of Holland (an auxiliary office of the KNMI) produces tidal rise predictions (Peeck *et al.*, 1982; Verlaan *et al.*, 2005). The SVSD is notified when the high tide at any reference station is expected to exceed the information level which is as much as 40 to 50 cm below the warning level. This message is generally issued about ten hours before the water is likely actually to reach that level. Based on the information supplied and on his own experience, the SVSD officer on duty a tidal hydrologist will decide whether or not it is expedient to staff the warning bureau, the SVSD action centre. This will normally be done whenever the warning or alarm level is expected to be reached or exceeded. The SVSD officer will issue warnings and/ or alarms. Wherever possible, this will be done at least 6 hours in advance of high tide, so that the dike and dam authorities have time to prepare. These warnings will be issued to a number of bodies concerned with the safety of the coastal provinces, including:

- water boards and dike and dam authorities
- Rijkswaterstaat field services
- the provincial public works authorities
- the Ministry of the Interior (Fire Service and Disaster Response Department).

As soon as an alarm is issued, announcements are broadcasted on radio and TV news bulletins.

2.4 Data Description

The sea water level, surge, atmospheric pressure and wind speed/direction time series data from seven coastal stations along the Dutch coast are monitored and provided by the Directie Noordzee (DNZ). Water levels are sampled at 0.0167 Hz and averaged over period of 10 minutes. Each time series that was made for us initially available begins January 1st, 1990 and ends on March 31st, 1996, which results in 337249 continuous samples in total for the 10 min times series data and 54768 for the averaged hourly times series (Table 2-1). The surge time series data is obtained by subtracting the observed water level with tide (astronomical forces) based on harmonic analysis, formulated as:

$$Surge = Water\ level\ (observed) - Tides \quad\quad\quad (2.1)$$

TABLE 2-1: DATA DESCRIPTION FROM TIDAL STATIONS IN THE DUTCH COAST (1990-1996).

Code	Station Name	Water levels				Surges		
		Max range	Avg. height	Sig. height	Var	Max range	Var	% diff
		[cm]	[cm]	[cm]	[cm^2 x10^3]	[cm]	[cm^2 x10^3]	
EPF	Euro platform	438	162.3	219.1	3.87	357	0.563	48.7
HvH	Hoek v. Holland	471	171.5	229.4	4.63	358	0.708	50.6
K13	K13 platform	468	156.4	208.8	2.68	332	0.773	46.6

TABLE 2-2: DATA SEPARATION FOR WATER LEVEL AND SURGE DATA INTO TRAINING, CROSS-VALIDATION AND TESTING DATA SETS FOR NON-STORMY AND STORMY CONDITIONS.

Time Index	Cross validation for optimizing model parameters				Model testing			
	Non-storm		Storm		Non-storm		Storm	
	Train	CV	Train	CV	Train	Test	Train	Test
Start	1	38200	1	35500	1	47473	1	43001
End	38199	38600	35499	35900	47472	49656	43000	45160

Figure 2-4 visualizes the relationship between air pressure, wind speed with direction of 120 degrees from North, water level and surge at Hoek van Holland. This study concentrates on predicting the water level and surge at the Hoek van Holland (HvH) tidal station at where the entrance of Rotterdam Port is located. The additional data from neighboring tidal stations, like Euro platform and K13, were also utilized.

Table 2-1 lists data description from these tidal stations. In order to evaluate the model performance during various conditions, the water level and surge time series data are divided into training, cross-validation (CV) and testing data sets for non-stormy and stormy conditions as listed in Table 2-2.

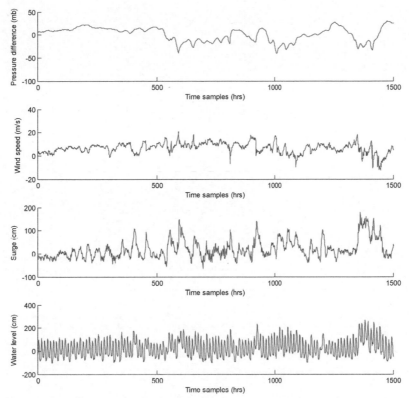

Figure 2-4: Relationships between the long-shore winds, surge, water level, and air pressure difference at Hoek van Holland.

2.5 Summary

The main case study is to predict the storm surges during non-stormy and stormy conditions along Dutch coast, specifically at Hoek van Holland as the important location (entrance) of Rotterdam Harbour. Some variables with 10 minutes and 1 hour resolutions are available, include: air pressure, wind, surge and water level. Maximum surge level and significant water level height in the time series used is 358 cm and 229cm, respectively.

CHAPTER 3: STORM SURGE MODELING

"Storm Surges Could Wreak $300 Billion Damages in 10 U.S. Coastal Cities."

Laura Mazzuca Toops

This chapter presents the fundamental principles of physical oceanography, tides, surge, spectrum-based wave model, several operational storm surge models for the North Sea.

3.1 Introduction

One of the most impressive natural phenomena is oceanic waves which have interesting behaviors from the chaotic motions due to a hurricane to a mild swell on a tropical beach. Many scientists are interested in the dynamics and kinematics of the waves: how they are generated by the wind, why they break and how they interact with currents and the sea bottom. Ocean and/or hydraulic engineers are involved in design, operation or management of structures or natural systems in the marine environment, such as dykes, storm surge barriers, beaches and ship navigation. One of the most important waves in ocean is storm surge that is generated due to a strong storm or hurricane. In shallow oceanic water, the structures and marine systems are much affected by storm surge. The knowledge on the physical behavior of this kind of waves is therefore required not only for designing dykes and storm surge barriers but also building a predictive storm surge model.

3.2 Physical Oceanography

3.2.1 Ocean waves and its classification

Ocean waves are mechanical waves that propagate along the interface between water and air; the restoring force is provided by gravity, and so they are often referred to as surface gravity waves (WMO, 1998; Stewart, 2002; Holthuijsen, 2007). As the wind blows, pressure and friction forces perturb the equilibrium of the ocean surface. These forces transfer energy from the air to the water, forming waves. In the case of monochromatic linear plane waves in deep water, particles near the surface move in circular paths, making ocean surface waves a combination of longitudinal (back and forth) and transverse (up and down) wave motions (Figure 3-1a).

The ideal ocean surface wave is sinusoidal with celerity (c) crests and the troughs having identical shapes and the wave having one fixed wavelength and orbital progressive, with water particles under the wave moving in orbital paths that make one complete cycle with the passage of one complete wave. On a spatial scale, the horizontal distance between two adjacent crests or troughs is defined as the wave length (L) and the vertical distance from the top of the crest to the bottom of the adjacent trough is defined as the wave height (H). On a temporal scale, the time that it takes for two consecutive crests to pass a fixed point is defined as the wave period (T). The inverse of the period is the wave frequency (f), which is a measure of the number of times one complete wave will occur per unit time (in cycles per second or hertz). Finally, the speed with which a wave crest moves horizontally across the ocean surface is defined as wave celerity (c) or phase speed (m/s).

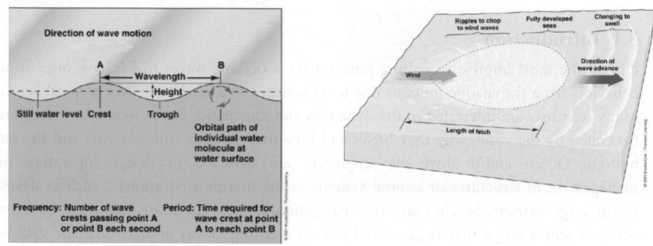

Figure 3-1: (a) Sinusoidal ocean wave form; (b) Wind generating sea and swell (Holthuijsen, 2007).

As the depth into the ocean increases, the radius of the circular motion decreases. By a depth equal to half the wavelength λ, the orbital movement has decayed nearly to zero. The speed of the surface wave is well approximated by:

$$c = \sqrt{\frac{g\lambda}{2\pi} \tanh\left(\frac{2\pi d}{\lambda}\right)} \qquad (3.1)$$

where c is phase speed, λ is wavelength, d is water depth and g is acceleration due to gravity at the Earth's surface.

In fact, the real ocean waves do not, of course, have a sinusoidal shape and rarely are found with a single wavelength or wave period. The ocean surface is quite chaotic and made up of many component waves of different periods and directions. The major driving force for ocean waves results from the wind forces on the ocean surface. Wind speed and wave activity are closely related. Besides the wind speed, there is the duration of storm and fetch. Fetch is the distance the wind blows over the water to generate waves. The wave speed is usually variable. Such variation produces waves of various sizes. The relationship between wind and waves is formalized in Beaufort scale.

The waves still under the action of the winds that created them are called *sea*. Whereas, the waves that have moved out of the generating area are known as swell. After, waves travelled a distance from the generating area, they have lost some energy (due to air resistance, internal friction, etc.) leading to a decrease in energy density. Thus, waves become lower in height. Seas usually have shorter periods and lengths and their surface appears more disturbed than for swells. Swells, being no longer under influence of wind, appear more orderly with well defined long crests and relatively long periods than seas (Figure 3-1b).

Ocean waves can be classified in at least into four classes based on the water depth, method of generation, wave period, and relationship to generating force) as described below.

3.2.1.1 Water depth

On the basis of water depth, the ocean waves can be categorized into:

- Deep water waves. Water too deep for waves to be affected by the seabed, typically taken as half the wavelength, or greater. Wave celerity is directly proportional to either wavelength or wave period. This means that waves with longer wavelengths (wave periods) will travel faster across the ocean surface.
- Shallow water waves. Typically this implies a water depth equivalent to less than one twentieth of wavelength. Wave celerity is directly proportional to depth. This means that as water depth decreases, waves slow down.
- Intermediate water waves. Transitional water waves between deep water waves and shallow water waves.

The processes that can affect a wave as it propagates into shallow water include: refraction, shoaling, dissipation due to friction, dissipation due to percolation, breaking, additional growth due to the wind, wave-current interaction and wave-wave interaction. Other wave transformation can occur due the presence of structures which interrupts wave propagation, e.g. diffraction and reflection. When waves propagate in shallow water, (where the depth is less than half the wavelength) the particle trajectories are compressed into ellipses. As the wave amplitude (height) increases, the particle paths no longer form closed orbits; rather, after the passage of each crest, particles are displaced a little forward from their previous positions, a phenomenon known as Stokes drift (Figure 3-2).

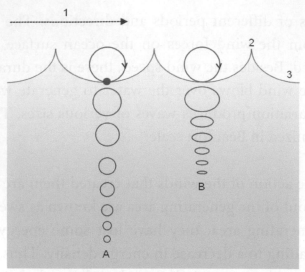

Figure 3-2: Deep water and shallow water with their water particle movements.

3.2.1.2 Method of waves generation

Ocean waves also may be classified by the method of their generation.

- Wind waves are generated when wind blows across the water surface and momentum is transferred from the wind to the water.
- Impact waves (such as Tsunamis) may be generated on the water surface by earthquakes or any other forms of impact (even, on a small scale, by a rock thrown into a pond).

3.2.1.3 Period of waves

Ocean waves can be classified by wave period as follows (see Figure 3-3):

- Ripples or capillary waves. The smallest waves have periods < 0.1 second and are generated by a small puff of wind and, because they are so small (are molecular waves), are restored by surface tension
- Gravity waves. The most common waves have periods between 1 sec and 30 seconds (with the most energy centered around 10 seconds), are generated by the wind and storms, and are restored by gravity.
- Long waves. Waves have periods greater than 5 min periods which are generated by intense storms (surges) and by earthquakes, and restored by gravity and the Coriolis force.
- Very long waves. The longest waves are the 12 hr and 24 hr tides, generated by the sun and moon and restored by bottom friction and the Coriolis force.

3.2.1.4 Relationship to the Generating Force

Some wind waves being actively generated (in an intense storm, for instance), as one will find out later, may be classified as free/forced waves.

- Free waves: ocean waves that run independent of their generating force (such as impact waves)

- Forced waved: waves that are dependent upon their generating force for their continued existence (such as the tides).

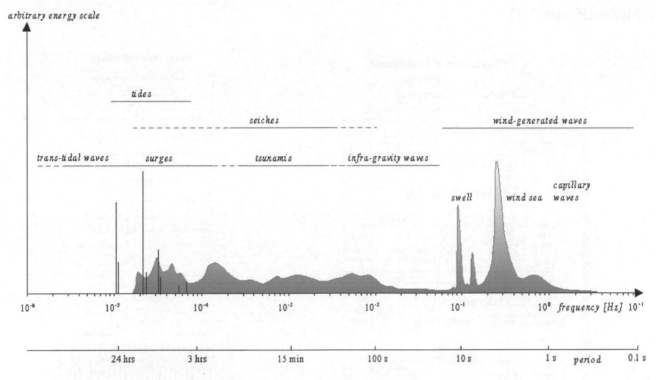

Figure 3-3: Wave categories based on the period of waves (Holthuijsen, 2007).

3.2.2 Tides

Tides are the cyclic rising and falling of Earth's ocean surface caused by the tidal forces of the Moon and the Sun acting on the oceans. Tides cause changes in the depth of the marine and estuarine water bodies and produce tidal currents. Prediction of tides is important for coastal navigation. The strip of seashore that is submerged at high tide and exposed at low tide, the inter-tidal zone, is an important ecological product of ocean tides (WMO, 1998; Holthuijsen, 2007).

The changing tide produced at a given location is the result of the changing positions of the Moon and Sun relative to the Earth coupled with the effects of Earth rotation and the local bathymetry. Sea level measured by coastal tide gauges may also be strongly affected by wind. More generally, tidal phenomena can occur in other systems besides the ocean, whenever a gravitational field that varies in time and space is present.

Tides may be semidiurnal (two high tides and two low tides each day), or diurnal (one tidal cycle per day). The various frequencies of astronomical forcing which contribute to tidal variations are called constituents. In most locations, the largest is the "principal lunar semidiurnal" constituent, also known as the M2 tidal constituent. Its period is about 12 hours and 24 minutes, exactly half a tidal lunar day, the average time separating one lunar zenith from the next, and thus the time required for the Earth to rotate once relative to the Moon (Figure 3-4).

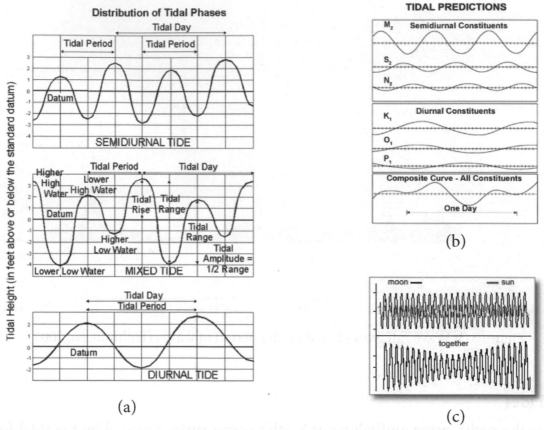

(a)   (c)

Figure 3-4: (a) Some types of tides: diurnal, semidiurnal and mixed; (b) and tidal harmonic constituents; (c) Moon and Sun forcing constituents creating tidal range variation: spring and neap tides (source: NOAA).

Around new and full moon when the Sun, Moon and Earth form a line, the tidal forces due to the Sun reinforce those of the Moon. The tide's range is maximum. This is called the spring tide. When the Moon is at first quarter or third quarter, the Sun and Moon are separated by 90° when viewed from the earth, and the forces due to the Sun partially cancel those of the Moon. At these points in the lunar cycle, the tide's range is minimum. This is called the neap tide. Spring tides result in high waters that are higher than average, low waters that are lower than average, slack water time that is shorter than average and stronger tidal currents than average. Neaps result in less extreme tidal conditions. There is about a seven day interval between springs and neaps.

3.3 Surges

Storm surge is simply water that is pushed toward the shore generated by combination forces of the wind friction and low air pressure around the storm (Figure 3-5). It is also defined as the oscillations of the water level in a coastal or inland water body in the period range of a few minutes to a few days, resulting from forcing from atmospheric weather systems (Gonnert *et al.*, 2001).

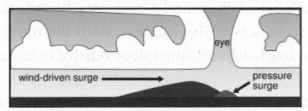

Figure 3-5: Storm surge driven by wind and air pressure.

In ocean, local wind waves can add to the water level, and the storm surge can be amplified (or reduced) by interference with the strictly regular astronomical tide. Extreme floods can be related to extreme storms, like cyclones or hurricane, which attack the open coast. In areas that are otherwise more sheltered (like the Baltic Sea), extreme floods can be generated by unusual sequences of wind set-up and air pressure variations. In addition, wind driven waves can be superimposed on the storm tide. This rise in water level can cause severe flooding in coastal areas, particularly when the storm tide coincides with the normal high tides (see Figure 3-6).

Irrespective of the weather, flood waves can be generated by distant, sub-sea earthquakes (such flood waves are called tsunamis), or, in arctic areas, by breaking glaciers. The effect also depends on the coastal profile. A gently sloping profile causes a high amplification of an approaching flood wave (and tidal wave) and a high set-up in response to local wind. A steep profile will tend to reflect the flood wave (and tidal wave), rather than amplifying it, and the locally generated set-up will be small if the water depth is large. Often, a shallow coastal profile occurs in places with low-lying lands. If so, the flood risk will be high.

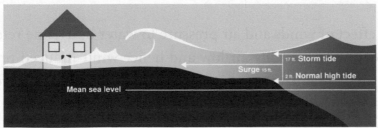

Figure 3-6 Storm surge flooding due to storm, tide, wave run-up and freshwater flooding (source: NOAA).

Coastal floods due to storm surges can be predicted with an accuracy that is largely determined by the accuracy of the meteorological predictions (which is within a couple of days). A major problem in this connection is that it is not possible to predict the route of a moving extreme depression like a cyclone. Analyzes of the general risk of coastal floods are made in connection with emergency preparedness planning, and also as a basis for design of coastal facilities and structures, such as flood embankments. The analyzes are difficult, because an 'observed flood' is not a single independent event in statistical terms. Rather, the flood is a consequence of a set of different determinants, like the tide, the wind and the air pressure, or of a set of sequences of these factors. Both the mean sea level and the flood height will vary along the coast, with a distinct slope relative to horizontal.

The mechanism leading potentially to coastal floods is well understood. Given the configuration of the coastline and the bathymetry, the severity of the storm surge depends primarily on wind speed, wind direction and duration. The meteorological conditions are affected by the path and the velocity of the depression systems, moving across the sea. When winds push water towards the coast, it tends to accumulate into what is commonly referred to as storm surge. If a particular high surge occurs together with a tidal maximum, both effects accumulate and serious flooding can result, depending on the coastal structure and their protection.

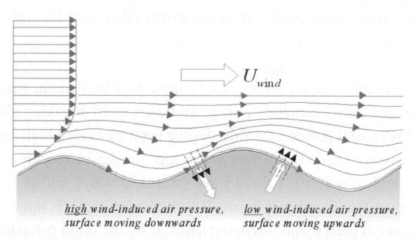

Figure 3-7: Wind-pressure variation induces wave (Holthuijsen, 2007).

To understand the effect of winds and air pressure to water body, one refers to (Miles, 1957) finding that the ocean waves modify airflow and hence wind-induced pressure at the water surface, such that they enhance their own growth. The waves are therefore generated by wind-induced surface pressure (Figure 3-7).

3.3.1 Tide-Surge Interaction

In North Sea region, it is well documented that the dominant semidiurnal M2 tide in some regions leads to a significant quarter-diurnal overtide (M4) due to nonlinear effects, namely, advection, wave drift, quadratic bottom friction, and time-dependent viscosity. The amplitude and phase of the M4 tide in the North Sea is mapped by models and compared with in situ sea-level measurements and also by satellite altimetry. The largest sea-level amplitudes, up to 25 cm, are found in the southern parts of the North Sea and in the English Channel. The M4 tide shows relatively rapid spatial variation in amplitude and phase and is localized to limited areas basically in shallow water. Possible transient generation of sea-level oscillations in the M4 band is due to interaction between the M2 tide and wind-generated current.

The swallow-water dynamical processes, which cause interaction between different tidal constituents as already demonstrated, also cause tidal and surge components of the sea levels and currents to interact. Suppose, for example that, there is a process which depends on the square of the total sea-level (Velickov, 2004):

$$\xi^2 = (T + S)^2 = T^2 + S^2 + 2TS \tag{3.2}$$

then the TS term in this case represents the interaction between the tides and the surges. In practice this interaction is difficult to describe in terms of analytical models and some knowledge can be gained from the numerical models. An alternative method is to analyze the distribution of the positive and negative surges relative to the high and low waters from the time series of the observations.

Tide-surge interaction on a local scale is very important because it is most apparent in shallow-water areas where large surges may be generated. The nonlinear interaction between the tides and surges may significantly change the design return period for coastal defences against flooding.

3.4 SWAN Wave Spectrum Model

Simulating Wave Nearshore (SWAN) model is a third-generation stand-alone (phase-averaged) wave model for the simulation of waves in waters of deep, intermediate and finite depth. It is also suitable for use as a wave hindcast model (Holthuijsen, 2007). However, SWAN can be used on any scale relevant for wind-generated surface gravity waves. The model is based on the wave action balance equation with sources and sinks. It adopts various grids (resolution, orientation, etc.), including nesting (only for uniform recti-linear grid).

The SWAN is a wave hindcast model based on spectrum analysis that can be explained as follows. Wind generated waves have irregular wave heights and periods, caused by the irregular nature of wind. Due to this irregular nature, the sea surface is continually varying. On the other hand, statistical properties of the surface, like average wave height, wave periods and directions, appear to vary slowly in time and space, compared to typical wave periods and wave lengths. The surface elevation of waves in the ocean, at any location and any time, can be seen as the sum of a large number of harmonic waves, each of which has been generated by turbulent wind in different places and times. They are therefore statistically independent in their origin. According to linear wave theory, they remain independent during their journey across the ocean. Under these conditions, the sea surface elevation on a time scale of one hundred characterstic wave periods is sufficiently well described as a stationary, Gaussian process. The sea surface elevation in one point as a function of time can be described as:

$$\eta(t) = \sum_i a_i \cos(\sigma_i t + \alpha_i)$$

(3.3)

where η is the sea surface elevation, a_i is the amplitude of the i-th wave component, σ_i is the relative radian or circular frequency of the i-th wave component in the presence of the ambient current and a_i is the random phase of the i-th wave component.

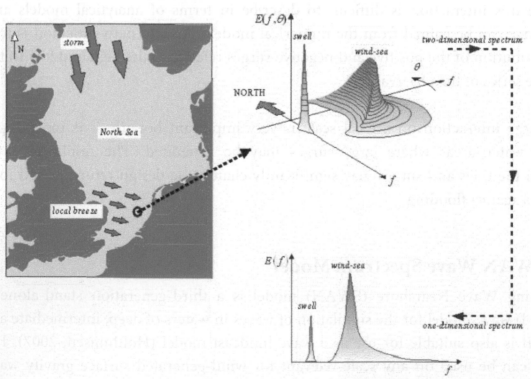

Figure 3-8: An interpretation of the wave spectrum of the Dutch coast adopted by SWAN model when a northerly swell, generated by a storm of the Norwegian coast meets a locally generated westerly wind sea (Holthuijsen, 2007).

Ocean waves are chaotic and a description in the time domain is rather limited. Alternatively, variance density spectrum is used, which is basically the Fourier transform of the auto-covariance function of the sea surface elevation. In SWAN, the energy density spectrum is generally used. On a larger scale the spectral energy density function becomes a function of space and time and wave dynamics should be considered to determine the evolution of the spectrum in space and time. An interpretation of the wave spectrum of the Dutch coast adopted by SWAN model is presented in Figure 3-8.

The other modeling application so-called Delft3D has presently integrated the SWAN model inside its numerical computation in order to enable an efficient and a direct coupling between e.g. circulation models (wave driven currents) and sediment transport models (stirring by wave breaking). Delft3D is developed by Delft Hydraulics for simulating two (either in the horizontal or a vertical plane) and three-dimensional flow, waves, water quality, ecology, sediment transport and bottom morphology and is capable of handling the interactions between those processes (Deltares, 2010).

3.5 Physcially-based Storm Surge Prediction Model

The basis of deterministic storm surge models are the shallow water equations, stating the physical laws of mass and momentum conservations (Dronkers, 1964), as follows:

$$\frac{\partial u}{\partial t} + u\frac{\partial u}{\partial x} + v\frac{\partial u}{\partial y} - 2\omega v + \frac{gu\sqrt{u^2+v^2}}{C^2(D+h)} + g\frac{\partial h}{\partial x} + \frac{1}{\rho}\frac{\partial p_a}{\partial x} - \frac{\rho_a C_d V_x \sqrt{V_x^2+V_y^2}}{\rho_w(D+h)} = 0 \tag{3.4}$$

$$\frac{\partial v}{\partial t} + u\frac{\partial v}{\partial x} + v\frac{\partial v}{\partial y} - 2\omega u + \frac{gv\sqrt{u^2+v^2}}{C^2(D+h)} + g\frac{\partial h}{\partial y} + \frac{1}{\rho}\frac{\partial p_a}{\partial y} - \frac{\rho_a C_d V_y \sqrt{V_x^2+V_y^2}}{\rho_w(D+h)} = 0 \tag{3.5}$$

$$\frac{\partial h}{\partial t} + \frac{\partial}{\partial x}\{(D+h)u\} + \frac{\partial}{\partial y}\{(D+h)v\} = 0 \tag{3.6}$$

where
t : time
x, y : space dimensions
h : water level elevation above reference level
u, v : depth averaged velocities in the x and y directions
C : Chezy coefficient
C_d : wind stress coefficient
D : depth below reference level
g : gravitational acceleration
p_a : atmospheric preassure

V_x, V_y : wind velocities in the x and y directions

ρ_a : density of the air

ρ_w : density of the water

ω : angular velocity of the earth

The perpendicular velocity component at closed boundaries is set equal to zero and the water level is given as a known time function at open boundaries. Due to the nonlinearity of the dynamics, the parallel velocity component vanishes at the boundaries in case of inflow.

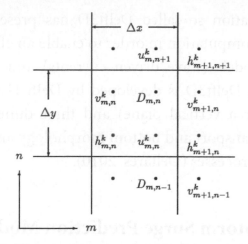

Figure 3-9: The computational grid of storm surge model.

In order to discretize these partial differential equations of the storm surge model, a space staggered grid is defined and finite difference scheme is employed, as depicted in Figure 3-9.

3.6 European Meteorological Offices and Storm Surge Models

3.6.1 North West Shelf Operational Oceanographic System (NOOS)

The operational storm surge prediction model has been used in different areas worldwide for many years. In the Netherlands, an automatic production line at the Royal Netherlands Meteorological Institute is used for that purpose. It contains a limited-area atmospheric model (called HIRLAM), the output of which drives a surge prediction model for which the above-mentioned Dutch Continental Shelf Model (DCSM) is used (Gerritsen *et al.*, 1995; Bode & Hardy, 1997; Verlaan *et al.*, 2005).

One of the large operational oceanography in Europe is North West Shelf Operational Oceanographic System (NOOS) which is developed by the EuroGOOS North West Shelf Task Team (NWSTT) in 2002 and operated by participating partners from the 9 countries bordering the extended North Sea and European North West Shelf (Belgium, Denmark, France, Germany, Ireland, Netherlands, Norway, Sweden, and UK), collaborating to develop and implement ocean observing systems for the North West Shelf area, with delivery of real time operational data products and services (Droppert, 2001). The role of NOOS is to create a design for a fully integrated observing and prediction system for the North Western part of the EuroGOOS area. This system has to meet the needs of a large number of groups in an effective way. Moreover it should be designed in such a way that all countries bordering the North West Shelf area are involved in the development and implementation of NOOS. The exchange of information through Internet has opened new opportunities for virtual networks where national networks, research vessels, cruises etc are being connected. Several institutions from European countries contribute on NOOS for providing North Sea storm surge predictions as listed below:

1. Koninklijk Nederlands Meteorologisch Instituut (KNMI)/ Rijksinstituut voor Kust en Zee (RIKZ) from Netherlands
2. Bundesamt für Seeschifffahrt und Hydrographie (BSH) from Germany
3. UK Meteorological Office (UKMO) from UK
4. Danish Meteorological Institute (DMI) from Danmark
5. Norwegian Meteorological Institute (DNMI) from Norwegia
6. Management Unit of the North Sea Mathematical Models (MUMM) from Belgium

The specifications of storm surge and circulation models of the European storm surge models within NOOS are described in Table 3-1.

3.6.2 KNMI and RIKZ

The Dutch institutions which are responsible on providing the North Sea storm surge predictions are Koninklijk Nederlands Meteorologisch Instituut (KNMI) and Rijksinstituut voor Kust en Zee (RIKZ). One storm surge model is so-called the WAQUA-in-Simona/DCSM98 storm surge model, which is developed by the National Institute for Coastal and Marine Management RIKZ, WL | Delft Hydraulics, the former Data Processing Division of Rijkswaterstaat and KNMI, is being used for day-to-day sea level predictions by KNMI's Maritime Meteorological Services since 1990.

The Dutch Continental Shelf Model (DCSM) is a large-scale water level prediction and storm surge prediction model that encloses the North West European shelf, including the British Isles. It has been developed by the National Institute of Coastal and Marine Management (RIKZ). In this model the shallow water equations with water quality and meteorological influences are described. The model is suitable for large scale transport calculations and for generating boundary conditions for nested models.

TABLE 3-1: STORM SURGE AND CIRCULATION MODELS (DE VRIES ET AL., 1995; PROCTOR, 1995; DROPPERT, 2001)

Country	Institution	Model Name	Model Area	Resolution	Type1	Characteristics
NL	KNMI in co-operation with RIKZ	WAQUA in SIMONA	Greater North Sea (Dutch coastal Shelf Model	8 km	Op.	2D Storm Surge Model with Data Assimilation (Kalman Filter)
Germany	BSH	BSHsmod BSHcmod	North Sea North Sea / Baltic Sea	6 nm 6 - 1nm 14 z levels	Op Op	2D Operational Storm Surge Model 3D Baroclinic Circulation Model
	IfM	HANSOM	North Sea / Baltic Sea	12' x 20" 14 z levels	Scient.	3D Baroclinic Circulation Model
Denmark	DMI	MIKE 21	North Sea / Baltic Sea	9 - 3 - 1nm	Op.	2D Storm Surge Model
	DHI	MIKE 3	North Sea / Baltic Sea	9 - 3 - 1nm z levels	Pre-op	3D Operational Baroclinic Circulation Model
Norwegia	DNMI	MI-POM	Greater North West European Shelf Nordic Seas, Norwegian Shelf	20 km 20 - 4 km	Op. Op.	2D Storm Surge Model 3D Baroclinic Circulation Model
	IMR	NORWEC OM	Greater North West European Shelf		Op.	3D Baroclinic Circulation Model with SPM and Ecosystem Model
	NERSC	DIADEM	North Atlantic and Nordic Seas	10-20 km North Sea	Pre-op.	3D isopycnic layered OGCM
		TOPAZ NSEA NWAG	Nordic Seas Atlantic Margin and Nordic Seas Atlantic Margin and Faroes	7 km 2 km	Pre-op. Pre-op. Pre-op.	3D hybrid OGCM 3D hybrid OGCM 3D hybrid OGCM
UK	UKMO	POL CS3 POL "model B" Shelf Model	North West European Shelf 48N to 63N 12W to 13E	1/9° lat by 1/6° long, 15 s levels	Op. Op. from June 2000	2D Storm Surge Model 3D Baroclinic Circulation Model. Surface forcing from mesoscale NWP. Deep ocean BCs from 1/3o FOAM climatology.
Belgium	MUMM / AWK	OPTOS-CSM	North Sea	2.5' x 2.5'	Op.	2D Storm Surge and Circulation Model
	Univ. of Liege	PCNOE	North Sea	10' x 10' 15 s levels	Scient.	3D Baroclinic Circulation Model with SPM and Ecosystem Model
B / UK	MUMM / NUE / POL / BODC	COHEREN S	North Sea	6' x 4' 20 levels	Scient.	Coupled 3D Baroclinic Hydrodynamic-Ecological Model

1 Op. = operational, Pre-op. = pre-operational, Semi-op. = semi-operational, Scient. = scientific

This page intentionally left blank

The model calculates the sea level and the depth averaged current on the Northwest European Continental Shelf on a grid with cells of approximately 8km x 8km (until September 1999 16km x 16km), using wind and pressure predictions from KNMI's limited area model HIRLAM. Kalman filter is used for real-time data assimilation of sea level observations. Sea level predictions are produced by the model 4 times per day for 48 hours ahead, closely following the available meteorological input. The structured curvilinear C grid type of Waqua and Triwaq used in the Dutch Continental Shelf Model (DCSM) is depicted in Figure 3-10.

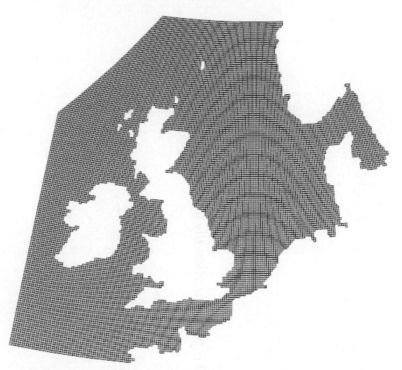

Figure 3-10: The structured curvilinear C grid type of Waqua and Triwaq used in the Dutch Continental Shelf Model (DCSM). The mesh resolution is approximately 8km (Ham, 2006).

Real-time assimilation of observed sea levels along the British and Dutch coasts gives a significant contribution to the high accuracy which is especially required for the management of the storm surge barriers in the Oosterschelde and the Rotterdam Waterway (Maeslantkering). Unlike in atmospheric modeling, however, it is not a vital part of sea level prediction.

Generally the model produces sea level predictions with a standard deviation of less than 15 cm along big parts of the Dutch and British coasts. For predictions less than 12 h ahead, the Kalman filter brings this even down to less than 10 cm.

It is difficult to quantify the performance of the model for extreme surges accurately, due to the fact that they are relatively rare. Every (severe) storm presents a new challenge for the models (and the meteorologists). Nevertheless, experience over the past few years has given confidence in the model results.

The Dutch Continental Shelf Model (DCSM) is run at KNMI. The schedules and processes of model execution until releasing a decision can be described as follows (see Figure 3-11):

- Prediction and observation data are stored in FTP server (NOOS) every time the model run is finished. A script runs collecting data and checking prediction data available every half an hour.

- Every half an hour the available prediction data are used to issue warning 6 hours or 12 hours ahead or even longer. The final decision should be made before that time.

- The way is to look at the clock, what time is the last prediction we are going to get before reaching 6 hours ahead. For example, there will be high water at 12:00PM noon (GMT time). A decision should be made at 06:00AM by looking at the final predictions available at 06:00AM, which is the midnight prediction. The midnight predictions approximately come out at 04:00AM. The four hours time is required for model run (about 2,5 hours) and reviewing the results. In general, the final predictions should be available at 06:00, 12:00, 18:00, 00:00 every day.

- There are several intermediate model runs at 03:00, 09:00, 15:00, 21:00 every day. At these times, new meteorological data are obtained and DCSM with data assimilation (EnKf) also runs.

- Dissemination of a warning takes about 30 minutes.

- DCSM (coarse scale) runs just a few minutes for 2 days ahead. A refined model for Rotterdam (detailed scale) takes about 4 hours. They depend on the number of points and time steps.

- For storing the data, the timeline is available in the database in which each value has analysis time (*anal_time*). The analyze time is used by meteorological agency. For example, if the data collected at midnight for a certain point may take 10 minutes and running the model takes 2,5 hours then *anal_time* for this case is at midnight since *anal_time* is the starting point of the model run regardless delays (computation, transmission). It is not the end of the model run.

- The analysis time depends on the data source (location) and have much different between data sources.

- Institutes of each country that have operational storm surge model predictions for the North Sea does not always provide all predictions to be stored in FTP server. Prediction data could be available once a day since they need time for re-evaluating the predictions and making a decision.

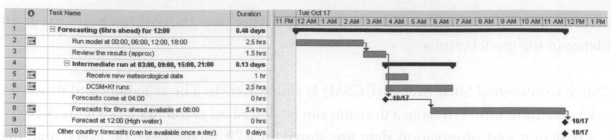

Figure 3-11: An example of operation schedule of the Dutch Continental Shelf Model.

3.6.3 European Centre for Medium-Range Weather Predictions (ECMWF)

The European Centre for Medium-Range Weather Predictions (ECMWF) was established by nineteen European States including the Netherlands in 1971 with the aim to produce weather predictions ten days ahead with the five-day predictions. The main objective of the ECMWF was to provide medium-range weather predictions and to carry out scientific and technical research directed towards the improvement of these predictions. Since then ECMWF has made operational prediction at least one ten-day prediction per day, and distributed it from its computer system to the systems of the national meteorological services of its Member States via a dedicated telecommunication network (ECMWF, 2007).

The ECMWF prediction system consists of five components: a general circulation model, an ocean wave model, a data assimilation system an ensemble prediction system (EPS). In 1998 a seasonal prediction system started to operate and in 2002 a monthly prediction system was introduced. A physical processes formulation in ECMWF is depicted in Figure 3-12.

The general circulation model is the first ECMWF numerical model which is a grid-point model with 15 levels up to 10 hPa, and horizontal resolution of 1.875 degrees of latitude and longitude, corresponding to a grid length of 200 km on a great circle. In April 1983 this grid-point model was replaced by a model with a spectral representation in the horizontal with a triangular truncation at wave-number 63. At a time when the spectral technique was introduced it was more accurate than the grid point model for the same computational cost. With increased resolution and the introduction of the semi-lagrangian technique, there is no longer any significant difference in accuracy between the two representations. After 1983, there are several evolutions of ECMWF model resolution. The last model resolution changes in 2006 with 799 spectral resolution and 91 vertical levels. In 1995 an explicit cloud scheme was introduced with clouds as prognostic parameters. It not only improved the cloud and precipitation predictions, but it had also a significant impact on the model dynamics, not only in the 10-day integration, but also on the preliminary fields for the analysis (Figure 3-12).

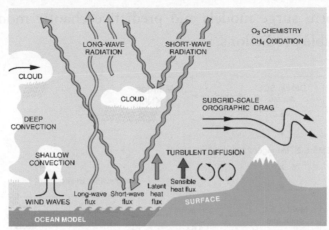

Figure 3-12: A physical processes formulation in ECMWF.

The ECMWF uses a global wave model plus a limited area model for the North Atlantic and the European waters. Currently, this wave model has been integrated into the atmospheric model allowing two-way interaction of wind and waves. It is now also incorporated in the monthly, seasonal and ensemble systems.

More advanced data assimilation techniques, like the variational data assimilation where the concept of a continuous feedback between observations and model are utilized in the view of increasing availability of synoptic data over the oceans. A four-dimensional system (4DVAR) has been implemented since 1997. The development of variational techniques has progressively allowed for a direct assimilation of satellite data, such as infrared and microwave sounder radiances, which impact on analyzed temperature and humidity fields.

The EPS simulates possible initial uncertainties by adding, to the original analysis, small perturbations within the limits of uncertainty of the analysis. From these alternative analyzes, a number of alternative predictions are produced. A wave model was included together with a crude allowance for the uncertainty of physical processes. In connection with tropical cyclones specially designed perturbations are created in the tropics.

3.7   Linking Predictive Chaotic Model with European Operational Storm Surge Models

One new contribution on predicting storm surges for the North Sea is to use a predictive chaotic model which is built based on the methods of nonlinear dynamics and chaos theory. A schematic diagram of the European storm surge models, their connections and predictive chaotic model – data-driven model (future component) are illustrated in Figure 3-14. The predictive chaotic model here is used in complementary with the existing European operational storm surge models and provide a fast-processing predictions for supporting decision-makers. A combiner or ensemble model is utilized to combine predictions from several operational storm surge models and predictive chaotic model in order to obtain more accurate and reliable predictions.

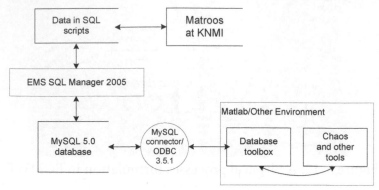

Figure 3-13: A schematic diagram of a connection between Matroos and other data-driven modeling environments.

Observation data from tidal stations in the North Sea and prediction data from European storm surge models are shared via FTP NOOS and stored in a database so-called Matroos. The processes of building a predictive chaotic model can be done automatically by accessing Matroos database (Figure 3-13). The predictions from predictive chaotic model can also be shared through this facility.

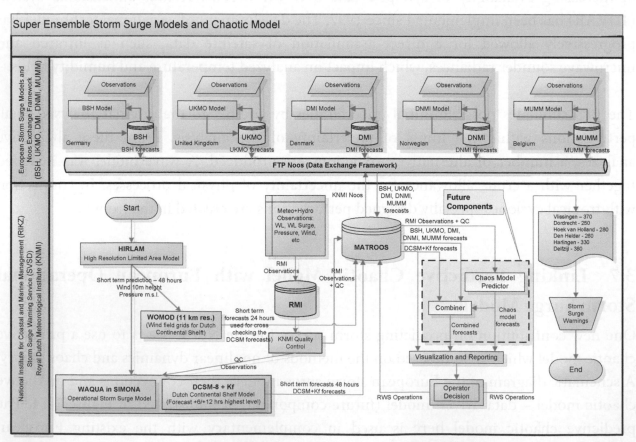

Figure 3-14: A schematic diagram of the European storm surge models, their connection and the future modeling components: the prediction combiner and chaotic model predictions.

3.8 Summary

The fundamentals of physical oceanography and coastal modeling have been discussed. Several physically-based numerical storm surge models for the North Sea have been operationally used. However, yet another technique of building predictive storm surge model based on data-driven techniques (computational intelligence and chaos theory) is introduced and presented in the following sections.

CHAPTER 4: COMPUTATIONAL INTELLIGENCE

"A computer would deserve to be called intelligent if it could deceive a human into believing that it was human."

Alan Turing

This chapter presents main instruments of the computational intelligence paradigm: artificial neural network, fuzzy system and evolutionary computation. They are the main tools in developing data-driven models and used in this work for building and optimizing the predictive storm surge model.

4.1   Introduction

For years sophisticated designs and beauty and of natural systems have been inspiration for scientists and engineers when they were thinking of the new approaches in building new artificial man-made systems. One of the lines of this thinking relates to the self-organising character of natural systems and the intellectual ability of some of them which is so tempting to mimic in artificial creations or at least in algorithms. Since 1940s this area of thinking is generally termed as Artificial Intelligence (AI). However, in the last two decades there was a certain shift, and a narrower "school of thought" has emerged – computational intelligence (CI) that is typically seen of consisting of: the three main approaches with the corresponding models: artificial neural networks (ANN), fuzzy systems (FS) and evolutionary computation (EC). AI can be seen as a more general multi-disciplinary research fields which include computer science, physiology, philosophy, sociology, physics and biology; its instruments relate to logic, deductive reasoning, expert systems, case-based reasoning and symbolic machine learning systems. CI is very closely to machine learning (ML), but also to soft computing (SC), data mining (DM), knowledge discovery in databases (KDD), intelligent data analysis (IDA). Of course, there are overlaps between all these areas, and other classifications and groupings of these areas and methods are possible.

For understanding the fundamentals of CI, one need to know what intelligence is. Intelligence has been defined as the abilities to think, understanding, comprehend, communicate, interpret, solve the problem and learn from experience. The extension of intelligence include: creativity, skill, consciousness, emotion and intuition. In the last century, scientists were attracted by a question "Can computer (machine) have intelligence?" This question caused more debate than the definitions of intelligence. In the middle of the last century, English scientist Alan Turing gave much thought to this question. He believed that machines could be created that would mimic the processes of the human brain. Turing strongly believed that there was nothing the brain could do that a well-designed computer could not. However, his statements are still visionary until now. Turing published his test of computer intelligence, referred to as the Turing test (Turing, 1950). The test consisted of a person asking questions via a keyboard to both a person and a computer. If the interrogator could not tell the computer apart from the human, the computer could be perceived as being intelligent.

TABLE 4-1: OVERLAPPING AREAS RELATED TO AI AND THEIR MAIN METHODS.

AI Sub Areas	Main Methods
Machine learning	□ Decision trees □ Model and regression trees □ Artificial neural networks (ANN) □ Bayesian learning, including belief networks □ Reinforcement learning (Q-learning) □ Statistical learning theory and support vector machines (SVM)
Soft computing (Methods tolerant for imprecision and uncertainty of data)	□ Fuzzy logic □ Artificial neural networks □ Evolutionary computing □ Probabilistic computing (including belief networks) □ Chaotic systems □ Parts of machine learning theory
Data mining (Preparation, reduction, finding new knowledge)	□ Automatic classification □ Pattern recognition (also called data analysis) □ Identification of trends (including statistical methods like ARIMA) □ Data normalisation, smoothing, data restoration □ Association rules and decision trees □ Neural networks □ Fuzzy systems □ Evolutionary and other global optimisation methods
Methods of non-linear dynamics	□ Chaos theory

In relation to the field of CI, data-driven modeling (DDM) can be seen as an approach to modeling that focuses on the use of CI methods in building models (often of natural systems) that would complement or even replace the "knowledge-driven" models describing behavior of physical systems (Solomatine, 2002). DDM uses methods developed in CI (Table 4-1) and tunes them for a particular application (Jang *et al.*, 1997).

The CI and machine learning techniques are the main source of methods for data-driven modeling, which are algorithms that estimate hitherto unknown mapping (or dependency) between a system's inputs and its outputs based on the available data (Mitchell, 1997). When such a dependency is discovered, it can be used to predict (or effectively deduce) the future system's outputs from the known input values (Figure 4-1). There are four main objectives of building DDM based on CI (machine learning) techniques, as follows (Mitchell, 1997):

- Classification –to find a way of classifying unseen examples on the basis of classified examples;
- Association – to identify association between features (which combinations of values are most frequent);
- Clustering – to discover groups of objects (examples) that are similar;
- Numeric prediction (regression) – to estimate the future state as a numeric (real) value based on training data.

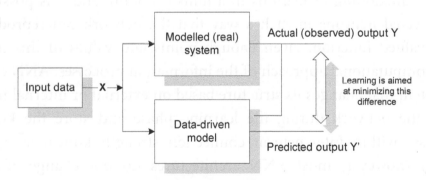

Figure 4-1: Learning process in data-driven modeling.

The CI techniques have been applied successfully to solve a number of real-world problems (Mitchell, 1997; Bishop, 2006; Engelbrecht, 2007; Haykin *et al.*, 2007; Hsieh, 2009). The development of hybrid model comprising several CI techniques currently becomes a trend, since none CI technique is superior to the others in all situations. By combining several CI techniques, one can take advantages on the respective strengths of the components of the hybrid CI system, and eliminate weaknesses of individual components (Figure 4-2).

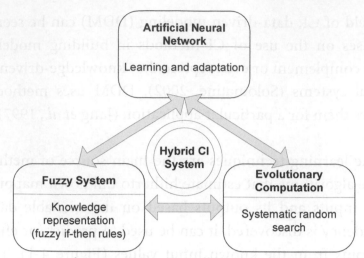

Figure 4-2: Main components of computational intelligence and their capability. Hybrid CI system combines the strength and eliminate the weaknesses of the individual CI component.

4.2   Artificial Neural Networks

Originally studied in the framework of AI, Artificial Neural Network (ANN) has become now one of the primary technologies in machine learning and data-driven modeling. ANN is a computational model that is inspired by the structure and functions of biological neural networks. Biological brain as the center of nervous system can be viewed as a complex, nonlinear and parallel computer. It has the ability to perform tasks such as pattern recognition, perception and motor control much faster than any computer. The ANN loosely imitates functioning of neurons in a human's brain, and it is possible to combine these interconnected neurons in such a way that the network can reproduce any multi-variable multi-valued function, given enough points and values of this function via the connectionist computational approach of the information processes. ANN can be treated as an adaptive system that changes its structure based on external or internal information that flows through the network during the learning phase and store the knowledge (from learning process) within inter-neuron connection strength known as synaptic weights (Haykin, 1999) (however, most ANN architectures do not change adaptively during operation). It has capabilities to learn, memorize and generalize the given information (data), and is characterized by distributed processing, adaptation and non-linearity. Therefore, it is commonly used to model complex relationships between inputs and outputs and to find patterns in data.

The ANN is increasingly being used in computing mimicking processes found in the nervous systems of vertebrates. The main characteristic of a biological neural network, (Figure 4-3, left), is that each neuron, or nerve cell, receives signals from many other neurons through its branching dendrites. The neuron produces an output signal that depends on the values of all the input signals and passes this output on to many other neurons along a branching fibre called an axon. In an ANN (Figure 4-3, right), input signals, such as signals from a television camera's image, fall on a layer of input nodes or computing units. Each of these nodes is linked to several other hidden nodes between the input and output nodes of the network. There may be several layers of hidden nodes, though for simplicity only one is shown here. Each hidden node performs a calculation on the signals reaching it and sends a corresponding output signal to other nodes. The final output is a highly processed version of the input.

The simplest neural networks relate an input signal to an output signal by means of a series of weighting functions that may involve a number of layers of interconnected nodes, including intermediate 'hidden layers'. Some applications have used additional filtering functions (essentially simply transfer functions) for each node in a hidden layer, so that the output will also depend on the form and parameterisation of these functions. A variety of techniques are available for determining the appropriate model structures and weights given a learning set of input and output data.

The architecture of a neural network is the specific arrangement and connections of the neurons that make up the network. One of the most common neural network architectures has three layers. The first layer is called the input layer and is the only layer exposed to external signals. The input layer transmits signals to the neurons in the next layer, which is called a hidden layer. The hidden layer extracts relevant features or patterns from the received signals. Those features or patterns that are considered important are then directed to the output layer, the final layer of the network. Sophisticated neural networks may have several hidden layers, feedback loops, and time-delay elements, which are designed to make the network as efficient as possible in discriminating relevant features or patterns from the input layer.

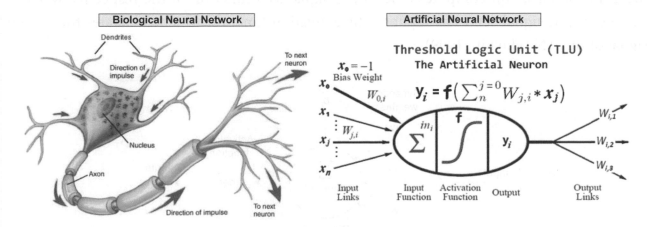

Figure 4-3: Schematic representation of ANN architecture and nervous system - modified (Rhode, 2011).

To mimic the way in which biological neurons reinforce certain axon-dendrite pathways, the connections between artificial neurons in a neural network are given adjustable connection weights, or measures of importance. When signals are received and processed by a node, they are multiplied by a weight, added up, and then transformed by a nonlinear function. The effect of the nonlinear function is to cause the sum of the input signals to approach some value, usually +1 or 0. If the signals entering the node add up to a positive number, the node sends an output signal that approaches +1 out along all of its connections, while if the signals add up to a negative value, the node sends a signal that approaches 0. This is similar to a simplified model of a how a biological neuron functions - the larger the input signal, the larger the output signal.

4.2.1 Mathematical model of artificial neuron

An artificial neuron (AN) is a model of a biological neuron which was firstly introduced by (McCulloch & Pitts, 1943). Since that, there have been developed hundreds of different models considered as ANNs. The differences are in the activation functions, topology and learning algorithms. Research on the applications of ANN has received great attention since the publication on back-propagating error learning algorithm by (Rumelhart *et al.*, 1986). This learning algorithm was firstly introduced by (Bryson & Ho, 1969) and further reported by (Werbos, 1974).

In principle, each AN receives signals from the environment, or other ANs, gathers these signals, and when fired, transmits a signal to all connected ANs. Figure 4-4 is a model of an artificial neuron. Input signals are inhibited or excited through negative and positive numerical weights associated with each connection to the AN. The firing of an AN and the strength of the exiting signal are controlled by an activation function. The AN collects all incoming signals, and computes a net input signal as a function of the respective weights. The net input signal serves as input to the activation function which calculates the output signal of the AN (Haykin, 1999).

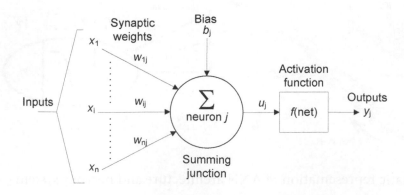

Figure 4-4: An artificial neuron model.

The inputs to such a AN may come from system causal variables or outputs of other ANs, depending on the layer that the AN is located in (Figure 4-4). These inputs form an input vector $\mathbf{x}=(x_1, ..., x_i, ..., x_n)$. The sequences of weights leading to the AN form a weight vector $\mathbf{w}_j=(w_1, ..., w_i, ..., w_n)$, where w_{ij} represents the connection weight from the i-th AN in this preceding layer to this AN j. The output of j or y_j is obtained by computing the value of function f with respect to the inner product of vector $\mathbf{x}$ and $\mathbf{w}_j$ minus $\mathbf{b}_j$, where $\mathbf{b}_j$ is the threshold value or called bias associated with this AN. The bias of the AN must be exceeded before it can be activated. The following equation defines the operation:

$$y_j = f(\mathbf{x_i}.\mathbf{w_j} - \mathbf{b}_j) \tag{4.1}$$

The function f is called an activation function. Its functional form determines the response of a AN to the total input signal. Several commonly used activation functions are as follows:

- Identity function: $f(\mathbf{x}_i, \mathbf{w}_i) = \mathbf{x}_i$
- Bias(Threshold) function: $f(\mathbf{x}_i, \mathbf{w}_i) = \mathbf{x}_i + \mathbf{w}_i$
- Linear function: $f(\mathbf{x}_i, \mathbf{w}_i) = \beta\mathbf{x}_i + \mathbf{w}_i$, where β is steepness parameter
- Sigmoid function: $f(\mathbf{x}_i, \mathbf{w}_i) = 1/[1+\exp(-\beta\mathbf{x}_i)]$
- Hyperbolic tangent: $f(\mathbf{x}_i, \mathbf{w}_i) = \tan(-\beta\mathbf{x}_i)$

An artificial neural network (ANN) is a layered network of ANs. An ANN may consist of an input layer, hidden layers and an output layer. ANs in one layer are connected, fully or partially, to the ANs in the next layer. Feedback connections to previous layers are also possible. Several different ANN types have been developed, for example:

- Single-layer ANNs, such as the Hopfield network;
- Multilayer feedforward NNs, including, for example, standard backpropagation, functional link and product unit networks;
- Temporal ANNs, such as the Elman and Jordan simple recurrent networks as well as time-delay neural networks;
- Self-organizing ANNs, such as the Kohonen self-organizing feature maps and the learning vector quantization;
- Combined supervised and unsupervised ANNs, e.g. some radial basis function networks.

These ANN types have been used for a wide range of applications, including diagnosis of diseases, speech recognition, data mining, composing music, image processing, prediction, robot control, credit approval, classification, pattern recognition, planning game strategies, compression, and many others.

4.2.2 Learning methods

An automated approach is required for determining the values of the weights w_i and the threshold θ. These values might be easy to calculate for simple problems. But suppose that no prior knowledge exists about the function. Only data is available. How can the w_i and θ values be computed? The answer is through learning. The ANN learns the best values for the w_i and θ from the given data. Learning process consists of adjusting weight and threshold values until a certain criterion (or several criteria) is (are) satisfied (Engelbrecht, 2007).

In order for an ANN to generate an output vector $\mathbf{y}_j=(y_1, ..., y_i, ..., y_n)$ that is as close as possible to the target vector $\mathbf{t}_j=(t_1, ..., t_i, ..., t_n)$, a training process, also called learning is employed to find optimal weight matrices $\mathbf{w}_i$ and bias vector $\mathbf{b}_j$, that minimize a predetermined error function that usually has the form:

$$E = \sum_P \sum_p (y_j - t_j)^2 \qquad (4.2)$$

where t_i is a component of the desired output t,

y_j is corresponding ANN output,

p is number of nodes,

P is number of training patterns.

Training a network is a procedure during which an ANN processes training set (input-output data pairs) repeatedly, changing the values of its weights, according to a predetermined algorithm, to improve its performance. Each pass through the training data is called epoch and the ANN learns through the overall change in weights accumulated over many epochs.

There are three main types of learning based on the presence or absence of the "teacher" and the information provided for the system to learn (Figure 4-5):

*   Supervised learning, where the neuron (or ANN) is provided with a data set consisting of input vectors and a target (desired output) associated with each input vector. This data set is referred to as the training set. The aim of supervised training is then to adjust the weight values such that the error between the real output, $y = f(\text{net}-\theta)$, of the neuron and the target output t, is minimized.
*   Unsupervised learning, where the aim is to discover patterns or features in the input data with no assistance from an external source. Many unsupervised learning algorithms basically perform clustering of the training patterns.
*   Reinforcement learning, where the aim is to reward the neuron (or parts of an ANN) for good performance, and to penalize the neuron for bad performance.

On the basis of the applied rules of learning, the learning techniques can be further classified into (Figure 4-5):

- Instance-based learning or lazy learning, where the generalization procedure is executed only when the new instance/pattern is presented, by searching similar or nearest distance to the training instances.
- Gradient descent learning, where the aim is to minimize the output error defined in terms of the weights and activation functions of the network.
- Stochastic learning, where the weights are adjusted in a probabilistic manner, for instance, simulated annealing with learning mechanism using Boltzmann machine.
- Hebbian learning, where the input-output pattern are characterized by adjusting a correlative weight matrix.
- Competitive learning, where the input pattern is presented, all neurons in the layer compete and the winning neuron undergoes weight adjustment. This strategy is called "winner-takes-all".

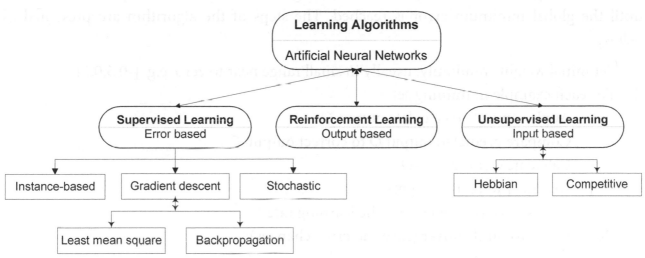

Figure 4-5: Classification of learning algorithms.

4.2.3 Multi-layer perceptron and back-propagation algorithm

The most common ANN model is the feed-forward multilayer perceptron (MLP). In feed-forward ANN, all signals flow in one direction from lower layers (input) to output layers (output). The MLP consists of a number of perceptrons which are arranged as a network structure. A perceptron is a basic model of ANNs which consists of a single layer ANN which its weights and biases can be trained to produce a correct target vector when presented with the corresponding input vector. Although a perceptron consists of input and output layers, it is essentially not a two layer network (Figure 4-6) because the output layer is the only real layer which consists of neurons performing the summation and non-linear activation transfer. The output computed by perceptron is +1 if the output is above a certain threshold and −1 otherwise.

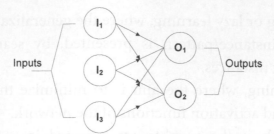

Figure 4-6: A single perceptron.

The learning algorithms for a perceptron can be of hill climbing or gradient descent search algorithms through weight space. The gradient descent search algorithm seeks the weight vector that minimizes the error by starting with an arbitrary initial weight vector, then repeatedly modifying it in small steps. At each step, the weight vector is altered in the direction that produces the steepest descent along the error surface. This process continues until the global minimum error is reached. The steps of the algorithm are presented as below:

i.   Set initial weights randomly, usually in small range near to zero, e.g. [-0.5,0.5]
ii.   For each example in training set
      -    Apply input, I_j to calculate output O.
      -    Compare predicted output O to correct output T.
      -    Calculate error: $e = T - O$
      -    Use error to revise weights:
           $w_j \leftarrow w_j + \eta I_j e$, where η is the learning rate.
iii. Repeat step (ii) until convergence (no error changes).

A number of perceptrons arranged into several layers (input, hidden and output) creates an MLP network architecture. The MLP is known as a supervised network because it requires a desired output in order to learn. The goal of this type of network is to create a model that correctly maps the input to the output using historical data so that the model can then be used to produce the output when the desired output is unknown. A graphical representation of an MLP is shown below.

The MLP and many other neural networks learn using an algorithm called back-propagation which is essentially a gradient descent technique that minimizes the network error function. With back-propagation, the input data is repeatedly presented to the neural network. With each presentation the output of the neural network is compared to the desired output and an error is computed. This error is then fed back (back-propagated) to the neural network and used to adjust the weights such that the error decreases with each iteration and the neural model gets closer and closer to producing the desired output. This process is known as "training".

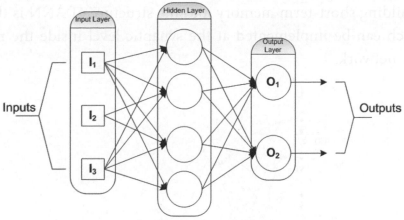

Figure 4-7: A feedfoward multi-layer perceptron.

Back Propagation involves minimization of an error function which is accomplished by performing gradient descent search on the error surface. With the conventional mean squared error function and with an additional term to dampen oscillations, the BP weight update rule can be shown to be:

$$\Delta w_{ji}(n+1) = \eta o_i \delta_j + \alpha \Delta w_{ji}(n)$$

where Δw_{ji} is the change in the weight connecting unit i to unit j, η is the learning rate a small positive constant, o_i is the output of unit i, α is the momentum term and the n indicates the epoch of the pattern presentation sequence. δ_j is the error associated with unit j commonly known as the instantaneous gradient and is calculated as:

$$\delta_j = \begin{cases} o_j'(t_j - o_j), \text{if unit } j \text{ is an output unit} \\ o_j' \sum w_{kj} \delta_k, \text{if unit } j \text{ is an hidden unit} \end{cases} \quad (4.3)$$

In this equation o_j' stands for the derivative of the output also called sigmoid prime and t_j is the target output of unit j. Assuming a sigmoid activation function for the units the derivative evaluates to:

$$o_j' = o_j(1 - o_j) \quad (4.4)$$

In practice, three aspects need to be considered during learning process of an ANN, include: choice of the training set and its size, selection of learning constants and stopping criteria.

4.2.4 Dynamic neural network

One type of neural network architectures is a dynamic neural network (Haykin, 1999; Bishop, 2006). For a neural network to be dynamic, it must be given memory that can be short-term and long-term memory depending on the retention time. Long-term memory is built into a neural network through supervised learning. However, if the problem has a temporal dimension, the short-term memory is needed to make the network dynamic. One simple way of building short-term memory into the structure of ANN is through the use of time delays, which can be implemented at the synaptic level inside the network or at the input layer of the network.

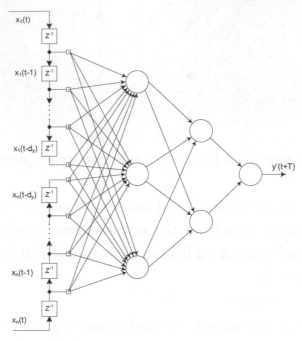

Figure 4-8: An architecture of dynamic neural network with time-delay inputs

4.3   Instance-Based Learning

Instance-based learning (IBL) (also called lazy learning) simply stores the presented training data and waits until a new query instance is given. Classifying the new query instance is performed by retrieving a set of similar related instances in the training data. In contrast, eager learning, the model or abstraction is built based on training data before receiving a new query instance. Nearest neighbor and locally weighted regression are two common approaches in the instance-based learning to approximating real-valued or discrete-valued target functions. The nearest neighbor approach represents instances as a point in Euclidean space whereas locally weighted regression constructs local approximation model (Aha *et al.*, 1991; Mitchell, 1997).

IBL requires less computational time for training but more time for predicting. The significant advantage is the capability to deal with very complex target function since the algorithm divides it into less complex local approximating functions. One disadvantage of instance-based approaches is the high computational cost on classifying new instances. An efficient indexing method for training data is a main solution in reducing the computation required at query time. The second disadvantage is that IBL typically considers all attributes of the instances when attempting to retrieve similar training examples. In case of the target depends on only a few of the many available attributes, the instances that are truly most similar may be a large distance apart.

4.3.1 k-nearest neighbors learning

The most common IBL algorithm is the k-nearest neighbors (k-NN) algorithm (Fix & Hodges, 1951). This algorithm assumes all instances correspond to points in the n-dimensional space $\Re^n$. The nearest neighbors of an instance are measured by an Euclidean distance. Let an arbitrary instance x be a feature vector [x^1, x^2, ..., x^n], where x^s denotes the value of the s-th attribute of instance x. Then the distance between two instances x_i and x_j is defined by:

$$d(x_i, x_j) = \sqrt{\sum_{s=1}^{n} (x_i^s - x_j^s)^2}$$

(4.5)

The k-NN algorithm can be used for either classification or regression problems. In classification task, Let a discrete-valued target function of the form $f: \Re^m \rightarrow V$, where V is the finite set [v_1, ..., v_s]. The value $f(x_q)$ is assigned by this algorithm as its estimate of $f(x_q)$ is the most common value of f among the k training examples nearest to x_q. If we choose $k=1$, then the 1-NN algorithm assigns to $f(x_q)$ the value $f(x_i)$ where x_i is the training instance nearest to x_q. For larger values of k, the algorithm assigns the most common value among the k nearest training examples.

```
Training algorithm:
     Store all training examples [x, f(x)]
Classification algorithm:
     Given a query instance xq to be classified,
     Let x1, ..., xk denote the k instances from the list of training
     examples
```

$$\text{Assign } \hat{f}(x_q) \leftarrow \arg\max_{v \in V} \sum_{i=1}^{k} \delta(v, f(x_i))$$

```
     where δ(a,b)=1 if a=b and where δ(a,b)=0 otherwise.
```

This algorithm can be used for regression by calculating the mean value of the k nearest training examples rather than calculate their most common value. For approximating a real-valued target function $f: \Re^n \rightarrow \Re$, the last statement of the above algorithm is replaced with:

$$\hat{f}(x_q) \leftarrow \frac{\sum_{i=1}^{k} f(x_i)}{k}$$

(4.6)

4.3.2 Distance weighted nearest neighbors algorithm

A useful enhancement of k-NN algorithm is to weight the contribution of each neighbor, so that the nearer neighbors to the query point x_q contribute more to the average than the more distant ones. A common weighting (interpolation) scheme is to give each neighbor a weight of $1/d$, where d is the distance to the neighbor, written as:

$$\hat{f}(x_q) \leftarrow \arg\max_{v \in V} \sum_{i=1}^{k} \delta(v, f(x_i)) \tag{4.7}$$

where

$$w_i \equiv \frac{1}{d(x_q, x_i)^2} \tag{4.8}$$

For real-valued target function, this formulation is replaced by:

$$\hat{f}(x_q) \leftarrow \frac{\sum_{i=1}^{k} w_i f(x_i)}{\sum_{i=1}^{k} w_i} \tag{4.9}$$

4.3.3 Locally weighted regression

The nearest-neighbor approaches described in the previous section can be thought of as approximating the target function $f(x)$ at the single query point $x = x_q$. Locally weighted regression is a generalization of this approach of k-NN algorithm in which it constructs an explicit approximation to f over a local region surrounding x_q instead of approximate the target function f at single query point $x=x_q$ (Cleveland, 1979). It uses nearby or distance-weighted training examples to form a local approximation to f. A linear function, a quadratic function, a multilayer neural network or some other functional form can be utilized for the approximating the target function in the neighborhood surrounding x_q. Let a linear function be the approximation for target function f, defined as:

$$\hat{f}(x_q) = w_0 + w_1 a_1(x) + \ldots + w_n a_n(x) \tag{4.10}$$

where $a_i(x)$ denotes the value of the i-th attribute of the instance x.

A gradient descent method can be used to find the coefficient $w_0 \ldots w_n$ so that the error of fitting linear function to a given training data is minimized. For global approximation to the target function, the weights that minimizes the squared error summed over the training data D, formulated as:

$$E \equiv \frac{1}{2} \sum_{x \in D} (f(x) - \hat{f}(x))^2 \tag{4.11}$$

with gradient descent rule:

$$\Delta w_j = \eta \sum_{x \in D} (f(x) - \hat{f}(x))^2 a_j(x) \qquad (4.12)$$

where η is a constant learning rate.

For a local approximation, the error criterion E must be reformulated to emphasize fitting the local training examples nearby query point x_q. Three error criteria to be minimized can be reformulated, as follows:

- The squared error over k-nearest neighbors

$$E_1(x_q) \equiv \frac{1}{2} \sum_{x \in k-neighbours} (f(x) - \hat{f}(x))^2 \qquad (4.13)$$

- The squared error over whole training data set D, but weighting the error of each training example by a decreasing function K of its distance to x_q

$$E_2(x_q) \equiv \frac{1}{2} \sum_{x \in D} (f(x) - \hat{f}(x))^2 K(d(x_q, x)) \qquad (4.14)$$

- A combination 1 and 2:

$$E_3(x_q) \equiv \frac{1}{2} \sum_{x \in k-neighbours} (f(x) - \hat{f}(x))^2 K(d(x_q, x)) \qquad (4.15)$$

There exists a broad range of alternative methods for distance weighting the training examples, and a range of methods for locally approximating the target function. In most cases, the target function is approximated by a constant, linear, quadratic or polynomial function.

4.4   Hierarchical Modular Models

A data-driven model may consist of several models, and one way of building it is to apply a hierarchical (tree-like) modular approach, leading to a decision tree or model tree. In classification, the so-called decision tree is an efficient, robust, and relatively simple model that is widely used (Breiman, 1984). In engineering most problems are regression (numerical prediciton) problems, and some researchers have attempted to use decision tree methods for numerical value prediction by dividing the range of values into small categories such as 0-3%, 4-6% etc, then using systems that build classification models. These methods often fail, because algorithms for building the decision trees cannot make use of the implicit ordering of such classes.

In the 1980-90s several learning algorithms were developed following the ideology of a decision tree. CART builds regression trees that differ from decision trees only in having values rather than classes at the leaves (Breiman, 1984) (in this respect this is a set of zero-order regression functions). MARS model constructs models which have splines basis functions (Friedman, 1991). We turned to the method developed by (Quinlan, 1992) and termed as the "model tree". The model tree is analogous to a set of piecewise linear functions, and in this sense can be seen as an extension of the CART method.

A typical example of model tree can be seen in Figure 4-9. One algorithm for inducing a model tree is M5 algorithm. This algorithm can learn efficiently and tackle tasks with very high dimensionality – up to hundreds of variables. Moreover, M5 model trees are generally much smaller than regression trees and have proven more accurate in the tasks investigated (Solomatine & Siek, 2006). Some advantages of M5 model trees (Wang & Witten, 1997; Witten & Frank, 2002) are that they are non black-box model, understandable, easy to use and to learn, fast on training and robust dealing with missing data.

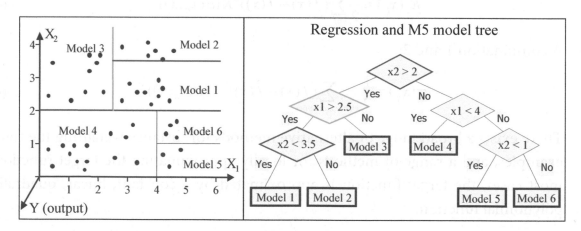

Figure 4-9: An example of M5 model tree and its hierarchically splitting of input-output space. Each local regions or data sets are approximated by local linear regression model (models 1 to 6 in the leaves)

The M5 model tree algorithm consists of three main procedures, as follows:

- Building the initial tree – an initial tree is constructed by a divide-and-conquer method with standard deviation reduction (SDR) splitting rule;
- Pruning the tree – a sub-tree is replaced with a leaf for avoiding over-fitting;
- Smoothing – it is used to compensate the sharp discontinuity between adjacent linear models at the leaves.

A model tree (MT), as mentioned previously, belongs to a class of committee machine which uses the 'hard' (i.e. yes-no) splits of input space into regions progressively narrowing the regions of the input space. Thus model tree is a or tree-like modular model which has splitting rules in non-terminal nodes and the expert models at the leaves of the tree. In M5 model trees the expert models are linear regression equation derived by fitting to the non-intersecting data subsets. Once the expert models are formed recursively in the leaves of the hierarchical tree, then prediction with the new input vector consists of the two steps: (i) classifying the input vector to one of the subspace by following the tree; and (ii) running the corresponding expert model. Brief description of model tree algorithm is presented below.

Assume we are given a set of N data pairs [$\mathbf{x}_i$, y_i], $i = 1,\ldots, n$, denoted by D. Here $\mathbf{x}$ is p dimensional input vector (i.e. x^1, x^1, $\ldots,x^p$) and y is target. Thus, a pair of input vector and target value constitutes the example, and the aim of the building model tree is to map the input vector to the corresponding target by generating linear regression equations at the leaves of the trees. The first step in building a model tree is to determine which input variable (often called attribute) is the best to split the training D. The splitting criterion (i.e. selection of the input variable and splitting value of the input variable) is based on treating the standard deviation of the target values that reach a node as a measure of the error at that node, and calculating the expected reduction in error as a result of testing each input variable at that node. The expected error reduction, which is called standard deviation reduction, SDR, is calculated by:

$$\text{SDR} = sd(T) - \sum_i \frac{|T_i|}{|T|} sd(T_i) \qquad (4.16)$$

where, T represents set of examples that reach the splitting node, T_1, $T_2,\ldots$, represents the subset of T that results from splitting the node according to the chosen input variable, sd represents standard deviation, $|T_i|/|T|$ is the weight that represents the fraction of the examples belonging to subset T_i.

After all possible splits are examined, M5 chooses the one that maximizes SDR. Splitting of the training examples is done recursively to the subsets. The splitting process terminates when the target values of all the examples that reach a node vary only slightly, or only a few instances remain. This division may produce models that overfit so that the tree must be pruned back, for instance by replacing a subtree (several models) with a leaf (one model). In the final stage, 'smoothing' is performed to compensate for the sharp discontinuities that will inevitably occur between the adjacent linear models at the leaves of the pruned tree. In smoothing, the outputs from adjacent linear equations are updated in such a way that their difference for the neighboring input vectors belonging to the different leaf models will be smaller. Details of the pruning and smoothing process can be found in (Wang & Witten, 1997; Witten & Frank, 2002).

4.5 Evolutionary and Other Randomized Search Algorithms

Evolutionary computation (EC) has as its objective mimicking processes from natural evolution, where the main concept is survival of the fittest: the weak must die. In natural evolution, survival is achieved through reproduction (Fraser, 1958). Offspring, reproduced from two parents (sometimes more than two), contain genetic material of both (or all) parents – hopefully the best characteristics of each parent. Those individuals that inherit bad characteristics are weak and lose the battle to survive. This is illustrated in some bird species where one hatchling manages to get more food, gets stronger, and at the end kicks out all its siblings from the nest to die. Evolutionary algorithms use a population of individuals, where an individual is referred to as a chromosome. A chromosome defines the characteristics of individuals in the population. Each characteristic is referred to as a gene. The value of a gene is referred to as an allele. For each generation, individuals compete to reproduce offspring (Holland, 1975).

Those individuals with the best survival capabilities have the best chance to reproduce. Offspring are generated by combining parts of the parents, a process referred to as crossover. Each individual in the population can also undergo mutation which alters some of the allele of the chromosome. The survival strength of an individual is measured using a fitness function which reflects the objectives and constraints of the problem to be solved. After each generation, individuals may undergo culling, or individuals may survive to the next generation (referred to as elitism). Additionally, behavioral characteristics (as encapsulated in phenotypes) can be used to influence the evolutionary process in two ways: phenotypes may influence genetic changes, and/or behavioral characteristics evolve separately.

Different classes of evolutionary algorithms (EA) have been developed (Crutchfield *et al.*, 2003; Engelbrecht, 2007; Dercole & Rinaldi, 2008):
- Genetic algorithms which model genetic-inspired evolution;
- Genetic programming which is based on genetic algorithms, but individuals are represented as trees;
- Evolutionary programming which is derived from the simulation of adaptive phenotypic evolution;
- Evolution strategies which are driven by modeling the strategy parameters that control variation in evolution;
- Differential evolution, which is similar to genetic algorithms, differing in the reproduction mechanism used;
- Cultural evolution which models the evolution of culture of a population and how the culture influences the genetic and phenotypic evolution of individuals;

- Co-evolution where individuals evolve through cooperation and competition to attain the necessary capability to survive.

Evolution via natural selection of a randomly chosen population of individuals can be thought of as a search through the space of possible chromosome values. In that sense, an evolutionary algorithm (EA) is a stochastic search for an optimal solution to a given problem. The evolutionary search process is influenced by the following main components of an EA:

- An encoding of solutions to the problem as a chromosome;
- A function to evaluate the fitness, or survival strength of individuals;
- Initialization of the initial population;
- Selection operators; and
- Reproduction operators.

Evolutionary computation has been used successfully in real-world applications, for example, data mining, combinatorial optimization, fault diagnosis, classification, clustering, scheduling, and time series approximation (Ashlock, 2006).

4.6   Summary

The main techniques in the computational intelligence paradigm: artificial neural network, fuzzy system and evolutionary computation have been utilized as the key instruments in developing data-driven models. and used in this work for building and optimizing the predictive storm surge model.

Nonlinear dynamical systems are common in real world though difficult to handle and understand. A number of examples of nonlinear systems which often exhibit complex chaotic behavior include: climate, hydrometeorology, economy, physics, physiology and socio-economy. Recent research in chaotic systems investigate fundamental properties of chaotic systems while computational intelligence technique is one of the examples and it becomes a general framework for modeling highly nonlinear dynamical systems.

CHAPTER 5: NONLINEAR DYNAMICS AND CHAOS THEORY

"No one welcomes chaos, but why crave stability and predictability?"
Hugh Mackay

This chapter describes the discovery of chaotic phenomena, methods of nonlinear dynamics and chaos theory, chaos in iterative maps and differential equations, properties of chaos, phase space reconstruction, and the ways to find the proper values of delay time and embedding dimension, stability of dynamics, prediction in chaotic system and recurrence plots.

5.1   Introduction

The idea of dynamical chaos was firstly introduced by Poincaré when he participated in a mathematical contest to model dynamically stable solar system (three body problem) by means of Newton's equations (Ivars, 1993). The three body problem consists of nine simultaneous differential equations which a solution in terms of invariants converges of these equations were to be found. Poincaré did not succeed in giving a complete solution, however his work was remarkable – it presented the key idea of chaos in the solar system - leading to the new development in the field of celestial mechanics. The idea of chaos with the principle of sensitive dependence on initial conditions was initiated by Poincaré in 1903 through his own words (Poincaré & Halsted, 1913; Poincaré, 1952):

> *"If we knew exactly the laws of nature and the situation of the universe at the initial moment, we could predict exactly the situation of that same universe at a succeeding moment. but even if it were the case that the natural laws had no longer any secret for us, we could still only know the initial situation approximately. If that enabled us to predict the succeeding situation with the same approximation, that is all we require, and we should say that the phenomenon had been predicted, that it is governed by laws. But it is not always so; it may happen that small differences in the initial*

conditions produce very great ones in the final phenomena. A small error in the former will produce an enormous error in the latter. Prediction becomes impossible, and we have the fortuitous phenomenon."

The concept of chaos was not deeply developed until 1960 when Edward Lorenz made a strange discovery. Lorenz worked on the issue of weather prediction at his MIT laboratory through focusing on a simplified set of equations that still retained some essential elements of the atmospheric system rather than studying the full equations describing atmospheric flow and weather phenomena. By means of the existing computer at that time, he tried to integrate the equations numerically in time and started a new simulation from the halfway result of a running calculation in order to obtain long-term predictions. When he plotted such a continued calculation together with the original calculation in one graph, he found that after a short period of time the two curves started to diverge rapidly, ending with completely different behavior. Subsequently, he figured out the reason of round-off effects resulting from the stored values that he used as initial conditions for the second calculations differed slightly from the original values. This led him to conclude that a tiny perturbation of the initial conditions can lead to enormous differences over time (Lorenz, 1963). The perspective of weather prediction provides an interesting metaphor to express the effect that small causes can have big impacts, well-known as Butterfly Effect (Glieck, 1987):

Does the flap of a butterfly's wings in Brazil set off a tornado in Texas?

With a modern PC today, one can easily retrace Lorenz' footsteps and understand the remarkably rich behavior of his simple system as well as the sensitive dependence on initial conditions.

5.2   Basics of Chaos

5.2.1 Dynamical system

Theory of dynamical systems tries to understand and describe the changes over time of the physical or artificial systems. Some examples of such systems are: the solar system, weather, motion of billiard balls, stock market, and so forth. Many areas of hydrometeorology, geophysics, economics and physiology involve a comprehensive analysis of the dynamical systems based on the particular laws governing their change. These laws are derived from a suitable theory such as Newtonian mechanics, fluid dynamics, mathematical economics (Tsonis, 1992; Strogatz, 2001).

All these models can be unified conceptually in the mathematical notion of a dynamical system, which consists of two parts: phase space and dynamics. The phase space is the collection of all possible states of a dynamical system. Each state represents a complete condition of the system at a certain moment in time. The dynamics is a rule that transforms one point in the phase space representing the current state of the system into another point representing the state of the system one time unit.

For example in planetary motion, a state can be the location and velocities of all planets and stars in the solar system and the dynamics is the laws of gravity which provide the position and masses of the planets and determine the forces acting on them. Once an initial state is chosen, the dynamics determines the state at all future times. If one want to know what the state will be two time steps ahead just apply the rule twice: one application gives us the state one unit time from now, and the second application gives the state one unit time after that, which is two units time from now (Hochman, 2011).

5.2.2 Phase space

A state of the system is defined as the value of the smallest vector such that at time t_0 it completely determines the behavior of a dynamical system for any time $t>t_0$. The components of the state vector are called state variables (Packard *et al.*, 1980; Tsonis, 1992; Abarbanel *et al.*, 1993). The evolution of a system can be visualized as a path in state space. A state space could be finite-dimensional consisting of an infinite number of points forming a smooth manifold, such as in ordinary differential equations and mappings. A set of differential equations can be used for describing a dynamical system. The collection of all possible states is called the phase space. Thus, the phase space is a subset of the state space. The dimension of phase space is the number of degrees of freedom of a dynamical system. In modeling perspective, it is a number of variables that completely describe the system. In the context of Hamiltonian systems, it is the number of pairs of state variables.

5.2.3 Various behaviors of dynamical system

The asymptotic behavior of a dynamical system can be classified into four types: equilibrium points, periodic solutions, quasi-periodic solutions or chaos (Ott *et al.*, 1994; Strogatz, 2001). An equilibrium point can be either stable (called sink) or unstable (called source). In stable equilibrium, all trajectories near sink (attractor) are moving towards it as time increases. A dynamical system has a periodic solution with a fixed period T if the trajectories of the dynamical system precisely return to itself. A certain period of T is the time needed to reach the same point in state space again, for example a limit cycle. In a quasi-periodic system, the period is not fixed and can be irrational, for instance a torus (settling down with repeated oscillation).

The last type of behavior is chaotic that was discovered by Lorenz (1963). A chaotic system is very sensitive to initial conditions and deterministic. The tiny distance between two points in state space can diverge exponentially as the system evolves. An attracting limit set is a set of stable asymptotic motions. An attractor is called strange attractor if the attracting limit set is chaotic. All chaotic attractors is strange, but not all strange attractor is chaotic.

5.2.4 Dynamical invariants

An invariant set S of a dynamical system has the property that every trajectory or orbit that begins in S remains in S (Abarbanel *et al.*, 1993; Scott, 2005). The trajectory and attractor are two examples of invariant set. An invariant set that is also a manifold is called an invariant manifold. Invariant manifolds provide a natural description of the dynamics close to an equilibrium or periodic orbit and can make it possible to work in lower dimensions than the phase space of the underlying system, since a smooth dynamical system restricted to an invariant manifold is itself a dynamical system. This reduction in dimension is at the central of manifold techniques. The sudden appearance or disappearance of attractors as parameters are varied can be investigated to understand the stability change of an invariant set that exists throughout the parameter region. Many invariant sets and manifolds have persistence properties under perturbations of the dynamical system.

5.2.5 Chaos in Iterative Maps

The essential aspects of chaos can be found in systems that are even more elementary. These are the so-called discrete maps or iterative maps (Verhulst, 1845; Strogatz, 2001; Jonker & van Reeuwijk, 2010). A one dimensional iterative map has the form:

$$x_{n+1} = f(x_n) \tag{5.1}$$

where n is non-negative integer number. Starting with an initial value x_0, one obtains a series by repeatedly applying the equation, i.e. $x_1 = f(x_0)$, $x_2 = f(x_1)$. As an example, consider the map $f(x) = rx$ that is sometimes used as a simple model for bacterial growth:

$$x_{n+1} = rx \tag{5.2}$$

In this equation x then represents the (normalized) number of bacteria at generation n and r represents the growth rate. The evolution of the time series for any r and x_0 can be predicted without problem. The value of x_n depends on r and x_0 via:

$$x_n = rx_0 \tag{5.3}$$

The long term behavior depends critically on r; for $n \to \infty$, $x_n \to 0$ if $|r| < 1$, $x_n/x_0 \to \infty$ for $|r| > 1$; only for $r = 1$ nothing happens, $x_n = x_0$.

The behavior of this linear mapping is not very exciting. As a model for bacterial growth it is also rather unrealistic to the extent that unbounded growth cannot persist owing to an inevitable shortage of food and mutual competition. These effects can be taken into account by adding an extra factor $(1-x)$ to the growth model. As soon as the (normalized) population x nears 1, this factor then reduces the effective growth rate. Thus taking $f(x) = rx(1-x)$ one obtains the so-called Verhulst-model (Nasell, 2001), also referred to as the logistic map:

$$x_{n+1} = rx(1-x), \quad r \in [0,4], \quad x0 \in [0,1] \tag{5.4}$$

The crucial aspect that makes the behavior so interesting resides in the nonlinearity of $f(x)$; the map has a quadratic term in x. Due to this nonlinearity it is very hard to find an analytical solution for x, in terms of r and x_0, as could be done for the linear mapping. For example, one can take different values of r and $0 < x_0 < 1$. Figure 5-1 depicts the series for four different values of r. Stationary, periodic and chaotic behaviors are observed.

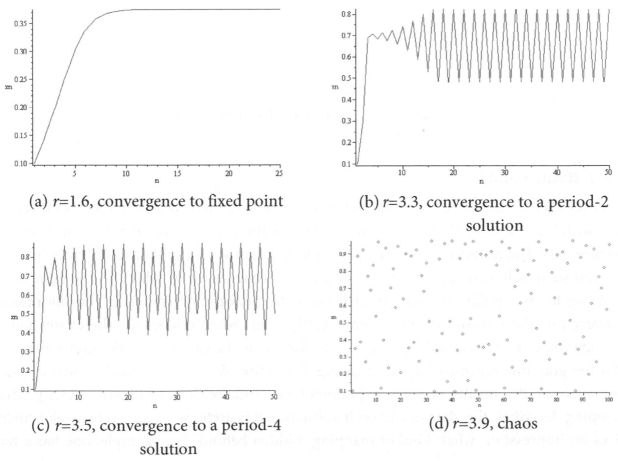

(a) $r = 1.6$, convergence to fixed point

(b) $r = 3.3$, convergence to a period-2 solution

(c) $r = 3.5$, convergence to a period-4 solution

(d) $r = 3.9$, chaos

Figure 5-1: A variety of behaviors of the logistic map for different values of parameter r.

5.3   Geometrical analysis of maps

5.3.1 Cobweb method

One of the interesting geometrical way to see how discrete maps is to use so-called cobweb plot (Figure 5-2). The principle of cobweb plot is as follows:

- Plot the mapping function $y=f(x)$ together with the line $y=x$;
- Use the graph $y=f(x)$ starting at xo to find the value of x by following the solid line;
- Utilize the line $y=x$, i.e. follow the horizontal line to the diagonal, then   follow the vertical dotted line giving you the point x_1 on the x-axis;
- Repeat the previous steps and find graphically $x_2 =f(x_i)$, then $x_3=f(x_2)$ and so on, until a fixed point is found.

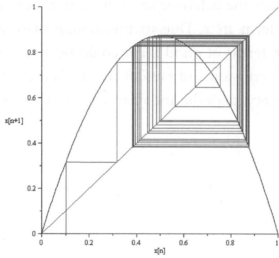

Figure 5-2: Cobweb plot of a logistic map.

5.3.2 Return plot

Another useful method to study the outcome of discrete maps is to use a so-called return-plot which are a plot of the value of x_{n+1} versus the previous value x for consecutive n (Figure 5-3). The number of n is typically set larger than some number in order to disregard the transient behavior, i.e. the initial part of the time series in which the system is still evolving to its equilibrium state (which could be periodic or chaotic). The information conveyed by the return-plots may seem slightly trivial when one knows the expression of the mapping that was used to generate the data, as is the case of logistic map it is easy to plot the gray line representing the mapping $f(x)$. However, in a practical situation where one analyses the results of some experiment one usually does not know exactly which mapping describes the datasets. In such a situation it is useful to make a return-plot since it gives an impression what kind of mapping hidden behind. For example, one has a hard time figuring out which data-set is uncorrelated noise, chaotic data from a one-dimensional mapping, or chaotic data from a higher-dimensional mapping (Strogatz, 2001).

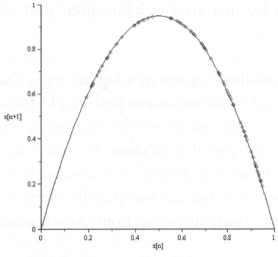

Figure 5-3: Return plot of a logistic map.

5.3.3 Fixed points and stability analysis

Some understanding of chaotic behavior is of importance. Figure 5-1(a) illustrates the stationary behavior and one notice in the system for $r<3$ and the series converges to a fixed point $x_n \rightarrow x^*$. A fixed point solution of the map $x_n \rightarrow x^*$ can be found analytically by:

$$\lim_{n \to 0} x_{n+1} = x_n \tag{5.5}$$

and hence the fixed point solution x must satisfy:

$$x^* = f(x^*) \tag{5.6}$$

For the logistic map, one can find two possibilities:

$$x^*=rx^*(1-x^*) \rightarrow x^*=0 \text{ or } x^*=1-1/r \tag{5.7}$$

Indeed, the series for $r=8/5$ depicted in Figure 5-1(a) converges to $x=3/8$. For any initial value $x \in <0, 1>$, the deviation will inevitably lead to $x^*=3/8$. Hence, one can say that for $r=8/5$ there is a fixed point $x^*=3/8$ which is stable and a fixed point $x^*=0$ which is unstable.

Stability of fixed points can be more precise understood as follows. If one start the sequence at a fixed point $x_0=x^*$, the system will remain in the fixed point.

5.4   Bifurcations

The possible asymptotic values (equilibria/fixed points or periodic orbits) as a function of a parameter in the dynamical system can be plotted in so-called bifurcation diagram. A fixed point or invariant point is a point that does not change or is mapped to itself by a function –

$f(x_0)=x_0$. For small values of the parameter, the dynamical system will be linear and a unique fixed point will exist. As the parameter is changed to ranges, where nonlinearity becomes important, instabilities in the form of new fixed points or solutions with qualitatively different dynamical behavior may arise at bifurcations (Ott *et al.*, 1994; Strogatz, 2001; Jonker & van Reeuwijk, 2010).

Figure 5-4 shows the bifurcation diagram of a logistic map. The fixed point solution, the period-2 solution, the period-4 solution and so forth can be found as the increased value of parameter *r*. The fixed point may lose its stability and split or bifurcate into two branches. Subsequently, for larger *r*, the period-2 solution becomes unstable itself and bifurcates into a period-4 solution and so on. This process of subsequent bifurcations is called period-doubling. With careful inspection, one can find that this behavior is no longer periodic after *r*=3.56, but becomes chaotic. The transition to/from chaos is often called intermittency.

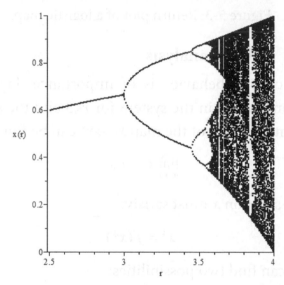

Figure 5-4: Bifurcation diagram of a logistic map.

5.5 Nonlinear Dynamics in Differential Equations

A dynamic system is a set of rules (mathematical formulae) that describes the time evolution of the system given some initial conditions and the time evolution is defined in some phase space $\Gamma \in \Re^d$. Such nonlinear systems can exhibit deterministic chaos (Hirsch *et al.*, 2004; Li, 2004), this is a natural starting point when irregularity is present in a signal or time series. Deterministic chaos comprises a class of signal intermediate between regular sinusoidal or quasi-periodic motions and unpredictable, truly stochastic behavior. Chaotic systems are treated as slightly predictable. These deterministic dynamical systems can be expressed by ordinary differential equations:

$$\frac{d}{dt}\mathbf{x}(t)=\mathbf{f}(\mathbf{x}(t)) \tag{5.8}$$

(for example Lorenz attractor, Figure 5-5b) or in discrete time $t=n\Delta t$ by maps of the form:

$$\mathbf{x}_{n+1}=\mathbf{f}(\mathbf{x}_n) \tag{5.9}$$

A time series can be considered as a sequence of observations $s_n=s(\mathbf{x}_n)$ performed with some measurement function $s(.)$. Since the sequence s_n in itself does not properly represent the phase space of the dynamical system, one has to employ some technique to unfold the multidimensional structure using the available data (Hegger *et al.*, 1999).

A sequence of points $\mathbf{x}(t)$ in the phase space, which represents a solution of the dynamic system is called a trajectory. If the trajectories in the phase space move to a single sub-space regardless of the initial conditions, but never cross each another, then they form a certain geometrical figure, which is called an attractor (see Figure 5-5b). Such dynamic systems are dissipative implying that the energy is not conserved. An attractor can be multi-dimensional and lie (is bounded) in an m-dimensional phase-space, but has a dimension less than m. For deterministic systems for which long-term predictability is feasible, and further exhibit periodic trajectories, the attractors are of integer dimension, for instance limit cycle. When the dynamic system is very sensitive to initial conditions, trajectories are quasi-periodic, but bounded in a m-dimensional phase space, the attractors have non-integer dimension or fractal dimension. Such attractors are called strange attractors. Those systems are usually characterized by a broadband power spectrum (no dominating periodic frequencies) and long-term predictability is not guaranteed.

The main reason for applying chaos theory for practical problems like surge prediction is the existence of methods permitting to predict the future positions of a chaotic system in the state space.

5.5.1 Sensitivity to initial conditions

The Lorenz equations (Lorenz, 1963) for describing atmospheric flow and weather phenomena are:

$$\begin{aligned}
\dot{x} &= \sigma(y-x) \\
\dot{y} &= rx - y - xz \\
\dot{z} &= xy - bz
\end{aligned} \tag{5.10}$$

where r, b and σ are parameters and $\dot{x}$ denotes dx/dt. The sensitivity to initial conditions can be nowadays easily shown by using a modern computer. A small perturbation to the initial conditions leads to enormous different in the outputs over time. For example, the initial condition is perturbed by $x(0)=2+\varepsilon$, where $\varepsilon=0.0001$. Figure 5-5(a) illustrates the small perturbed initial condition in Lorenz model creates a huge difference in the outputs starting at $t=[15,20]$. In phase space, the initial perturbation can results in completely different and distant trajectories and flow directions Figure 5-5(b).

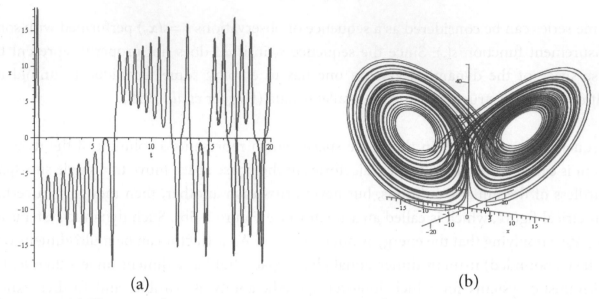

(a)   (b)

Figure 5-5: (a) The original Lorenz equation output in time-domain series $x(t)$ (blue) and the one with perturbed initial condition (red); (b) Phase space reconstruction in three dimensional space. This shows that a very small perturbation in the initial conditions leads to enormous difference in the output over time.

The trajectories in phase-space appear irregular, with no convergence to some sort of systematic, repetitive, pat tern. In addition, the computed solutions depend strongly on the initial conditions. A slight perturbation of the initial conditions, however small, gave rise to trajectories that sooner or later had nothing to do with the trajectories of the unperturbed system. This is the essence of chaotic systems. Because in any physical system one cannot know exactly the initial conditions, chaotic systems are in essence unpredictable in the long run. How long it takes for the system to reach this so-called 'prediction-horizon' depends: (a) on the accuracy with which one knows the initial conditions, (b) on the accuracy of the computer on which the calculation is performed and (c) on the intrinsic dynamics of the system itself.

5.5.2 Properties of chaos

The followings are some properties of the chaos dynamical systems:
- *Determinism*: The future dynamics are well defined by their initial conditions and predictable, and there are no random elements involved.

- *Small number of variables*: The dynamical systems can be represented by small number of variables, e.g. $x(n+1)=f(x(n), x(n-1), x(n-2))$.
- *Complex behavior*: The outputs of dynamical systems are complex and mimics random behavior.
- *Low dimensional in phase space*: The smooth trajectories and attractor of the dynamical systems can be appeared using a low dimensional phase space.
- *Sensitivity to initial condition*: Nearly identical initial values give very different final values, resulting it is not predictable in long run.
- *Bifurcations*: A small change in a parameter of chaotic system will generate a sudden change of the output pattern (see Figure 5-4).
- *Strange attractor*: The attractor of a chaos dynamical system is in low dimensional phase space and fractal (Figure 5-5b).

5.6 Phase Space Reconstruction – Method of Time Delay

When finding the structure of a dynamical system, one must reconstruct or embed time series in higher dimensional phase space. The most important phase space reconstruction technique is the method of delays. The method is known as Taken's embedding theorem (Takens, 1981) states that an embedding of a realization of the manifold M to space Γ exists if the dimension d of M is such that $d \geq 2m+1$. A manifold is any smooth (non-intersecting trajectory) geometrical space, for instance line or surface of a sphere. This reconstruction preserves the properties of the dynamical system which do not change under smooth coordinate/manifold adjustment, but it does not maintain the geometric shape of structures in phase space.

In principle, vectors in a new space, the embedding space are formed from time-delayed values of the scalar measurements. According to Taken's theorem, the dynamics of a time series $(x_1, x_2, \ldots, x_N)$ are fully captured or embedded in the m-dimensional phase space ($m>d$, where d is the dimension of the attractor) defined by the delay vectors:

$$Y_t = \left\{ x_t, x_{t-\tau}, x_{t-2\tau}, \ldots, x_{t-(m-1)\tau} \right\} \tag{5.11}$$

where τ is the delay time. A sequence of points in phase space is so-called trajectory. The time evolution of trajectories in most dynamical systems in nature tends to be dissipative and settled down to attracting points, so-called attractors. In practical applications, the delay time τ needs to be appropriately chosen in order to fully capture the structure of the attractor. This can be achieved by embedding the attractor in a smooth manifold. The lowest possible dimension of such manifold is called an embedding dimension.

5.7 Finding appropriate time delay

In real applications, the delay time τ needs to be appropriately chosen in order to fully capture the structure of the attractor. If τ is too small then the delay vectors are not independent, such that all points are accumulated around the bisector of the embedding space, resulting in loss of characteristics on the attractor structure. If τ is very large (i.e. much larger than the decorrelation time of the system), the different coordinates (delay vectors) may be almost dynamically uncorrelated. The straightforward choice of τ is usually made with the help of the zero-crossing autocorrelation function. The data are no longer correlated when the autocorrelation drops to zero. Further positive fluctuations can be interpreted as noise.

Tsonis & Elsner (1988) suggested that the time delay may be chosen as the lag time at which autocorrelation function falls below a threshold value which is commonly defined as $1/e$, especially if the autocorrelation function exhibit exponential decay. If data are suspected to be very noisy, τ has to be larger than the time when the normalized autocorrelation function decays to $1 - \sigma_{noise}^2 / \sigma_{signal}^2$. However, it must be pointed out that the autocorrelation function exploits the linear structures in the data.

Fraser & Swinney (1986) suggested to use mutual information and to choose τ such that it would correspond to the first minimum of the time delayed mutual information, and this approach demonstrated good performance in practical applications. The delayed mutual information is based on the Shanon's entropy and can be computed as follows: Given a time series of observable s, one can calculate the transitional probabilities $P_s(s_i)$ that a measurement s yields s_i. The information entropy is thus defined as:

$$H(s) = -\sum_{i=1}^{N} P_s(s_i) \log P_s(s_i) \tag{5.12}$$

The Shanon's entropy is a measure of the uncertainty associated with the measurement s. In other words, one can think of the degree of surprise when one reads the value of the measurement s. Low-probability (unexpected) measurements carry greater entropy than the high-probability (expected) measurements. The question now is how the value of the measurement $x(t+\tau)$ depends on $x(t)$ as a function of the time delay τ. If one denotes $s=x(t)$ and $q=x(t+\tau)$, then the conditional entropy can be written as:

$$H(q,s_i) = -\sum_{j=1}^{N} \left(\frac{P_{sq}(s_i,q_j)}{P_s(s_i)} \right) \log \left(\frac{P_{sq}(s_i,q_j)}{P_s(s_i)} \right) \tag{5.13}$$

where $P_{sq}(s_i,q_j)$ is the probability that measurements of s and q yields s_i and q_j. In this case one could define $H(q,s_i)$ as the uncertainty of q, given s_i. The mutual information is then defined as the amount that a measurement of $s=s_i$ reduces the uncertainty of q:

$$I(q, s_i) = H(s_i) + H(q) - H(q, s_i) \tag{5.14}$$

If the time delay is chosen to coincide with the first minimum of the mutual information, than the reconstructed state vector Y_t will consist of delay components that possess minimal mutual information between them. The mutual information method is probably the most comprehensive method of determining proper time delays when reconstructing the dynamics of the systems from time observables. The only drawback of this method is that requires a large amount of data and it is computationally expensive.

5.8 Estimating embedding dimension

5.8.1 Self-similarity: Dimension

A dynamical system is considered to be fractal if it contains similar structures at all length scales, known as self-similarity. The attractor of deterministic chaotic systems can exhibit an unusual kind of self-similarity and show structure on all length scales. A proper embedding dimension has to be searched such that the structure of the attractor becomes invariant. Invariant means not sensitive to the small perturbation of initial conditions. According to Whitney (1936), any smooth manifold of dimension d can be smoothly embedded in $m=2d+1$ dimension. According to the embedding theorem of Takens, a d-dimensional attractor can be embedded into a $(m=2d+1)$-dimensional phase space to estimate and describe the characteristics of the dynamic system. Sauer *et al.* (1991) further discussed the generalization of the embedding theorem, emphasizing the importance of the fractal dimension of the attractor for estimation of the minimal dimension of the embedding space, i.e. $m>2d$. Some authors (see, for example, Abrabanel et al., 1991) suggest that, in practice, $m>d$ would be sufficient.

The most widely used fractal dimension quantifier is the correlation dimension d_c, which is based on the correlation integral or function analysis (Grassberger & Procaccia, 1983b). This algorithm uses the phase space reconstruction from a scalar time series using the method of delays, where the reconstruction procedure involves the choice of time delay τ. Obtaining a non-integer, finite d_c for a time series demonstrates fractal scaling and indicates possible chaotic dynamics. The correlation sum for a collection of points Y_t in some vector space is the fraction of all possible pairs of points which are closer than a given distance r in a particular norm, Figure 5-6(a).

$$C(r) = \frac{1}{N_{ref}} \sum_{i=1}^{N_{ref}} \frac{1}{N} \sum_{j=1}^{N} H\left(r - |Y_i - Y_j|\right) \tag{5.15}$$

where H is the Heaviside step function, $H(y)=1$ for $y>0$ and $H(y)=0$ for $y \leq 0$, r is the radius of the sphere centered on Y_i, N is the number of points in Y_t, and N_{ref} is a calibrated number of reference points taken from Y_t that are needed to yield consistent statistics. Theiler's window can be utilized to exclude the points which are temporally correlated (Theiler, 1990). The norm $|Y_i - Y_t|$ is the standard Euclidean norm. The sum just counts the pairs (Y_i, Y_j) whose distance is smaller than r, or put in other words the relative frequency with which a typical trajectory enters the i-th volume element (sphere). Correlation function $C(r)$ is estimated for the range of r available from the time series and for several embedding dimensions m. Then $C(m,r)$ is inspected for the signatures of self-similarity, usually by estimating the slope of $Log\ C(r)$ versus $Log\ r$ plot. If the time series is characterized by an attractor, then for positive values of r, the correlation integral $C(r)$ is scales to the radius r by the power low:

$$C(r) \cong \alpha\, r^{\nu} \qquad\qquad (5.16)$$

where ν is called correlation exponent (slope of the $Log\ C(r)$ versus $Log\ r$ plot) and α is a constant. The slope can be generally estimated by the least-squares fit of a straight line over a certain range (length scales) of r, known as the *scaling region* (Figure 5-6).

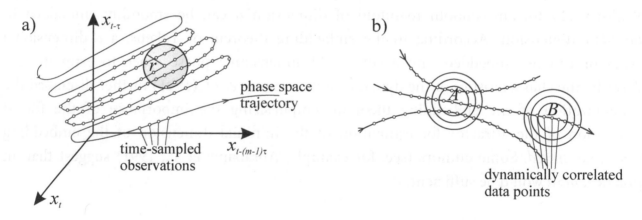

Figure 5-6: (a) Evolution of dynamic system in phase space showing the time-sampled data points and the neighborhood of the sphere in the correlation integral analysis (b) Influence of the temporal correlation on correlation integral analysis. While for point A there are some dynamically uncorrected neighboring points (lying on different trajectories), all neighboring points for point B are temporally correlated and thus stimulate a correlation dimension close to 1 (Velickov, 2004).

For a random process, ν varies linearly with increase of m, without reaching a saturation value, whereas for deterministic process, the value of the correlation exponent ν saturates and becomes independent of m for increasing embedded dimension. The saturation value d_c is defined as the correlation dimension of the attractor of the time series. If the correlation dimension d_c leads to a finite integer value, the underlying dynamics of the system is considered to be dominated by a strong periodic determinism. If the value of d_c is fractal and usually small then the systems is thought to be dominated by a low-dimensional deterministic chaotic dynamics governed by the geometrical and dynamical properties of an attractor.

5.8.2 False nearest neighbors

Another way to estimate the optimal value of m is to search for *false nearest neighbors* (FNN) in phase space at a given value of m (Kennel *et al.*, 1992). Consider a situation that m-dimensional delay reconstruction is the embedding, but an $(m-1)$-dimensional is not. The question is what happens when passing from m to m-1. This simply delineates along one coordinate and thus maps different parts of the attractor onto each other. When selecting a number of close points from such a region of the $\Re^{m-1}$, the images of the points will form different groups, depending on which part of the attractor the points are sampled. This lack of a unique location of all the images in m-1 dimensions is reflected by finding false neighbors, meaning that the determinism is violated. When increasing m, starting from small values of the embedding, one can thus detect the minimal (optimal) embedding dimension by finding no more false neighbors. The idea of the FNN algorithm is the following. For each point s_i in the time series look for its nearest neighbors s_j in a m-dimensional space, calculate the distance $\|s_i - s_j\|$, iterate both points and compute:

$$R_i = \frac{\left| s_{i+1} - s_{j+1} \right|}{\left\| s_i^{'} - s_j^{'} \right\|}$$ (5.17)

If R_i exceeds a given heuristic threshold R_t, this point is marked as having a false nearest neighbor. The criterion that the embedding dimension is high enough is that the fraction of points for which $R_i > R_t$ is zero, or at least sufficiently small. In a presence of noise, one should not expect drop of the percentage of false neighbors to zero in any dimension. Furthermore, if time series in question is stochastic, there will not be a substantial drop of the false neighbors with the increase of the embedding dimension.

5.8.3 Cao's method

The FNN algorithm has a drawback associated with the subjective choice of the threshold in order to ensure a correct distinction between low-dimensional chaotic data and noise. Different time series data may have different threshold values. These imply that it is very difficult and even impossible to give an appropriate and reasonable threshold value which is independent of the dimension m and each trajectory point, as well as the considered time series data. To avoid this, Cao proposed a modified algorithm, sometimes called the averaged false neighbors (AFN) method (Cao, 1997). Cao's approach is based on the estimation of two parameters $E1$ and E^* which are basically derived from the quantities defined by the FNN method. Based on the construction of the time delay vectors from the time series $x_1, x_2, ..., x_N$ an m-dimensional vector is defined by $y_i(m)=(x_i, x_{i+\tau}, x_{i+2\tau}, ..., x_{i+(m-1)\tau})$, where $i =1,2,...,N-(m-1)\tau$ and τ is the time delay. Similarly to the FNN method, the AFN approach defines the quantity:

$$a(i,m) = \frac{\left\| y_i(m+1) - y_{n(i,m)}(m+1) \right\|}{\left\| y_i(m) - y_{n(i,m)}(m) \right\|} \qquad (5.18)$$

where $\|.\|$ is the maximum norm, $y_i(m+1)$ is the i-th reconstructed vector for embedding dimension m and $n(i,m)$ is an integer such that the m-dimensional time-delay vector $y_{n(i,m)}(m)$ is the nearest neighbor of $y_i(m)$. Subsequently, the quantity of $E1$ is formulated as the mean value of all FNN distance $a(i,m)$:

$$E(m) = \frac{1}{N - m\tau} \sum_{i=1}^{N-m\tau} a(i,m) \qquad (5.19)$$

The $E(m)$ depends only on the dimension m and the time delay τ. The variation from m to $m+1$ can be investigated by $E1(m)=E(m+1)/E(m)$. The $E1(m)$ stops changing when m is greater than some value m_0 if the time series comes from an attractor. Then m_0+1 is the minimum embedding dimension to be obtained. It is necessary to define another quantity E^* which is useful to distinguish deterministic from stochastic time series, formulated as:

$$E^*(m) = \frac{1}{N - m\tau} \sum_{i=1}^{N-m\tau} \left| x_{i+m\tau} - x_{n(i,m)+m\tau} \right| \qquad (5.20)$$

These quantities are computed for different, progressively increasing values of the embedding dimension m. Subsequently, the global behaviors of $E1$ and E^* as functions of dimension m are respectively used for estimating the minimum embedding dimension and determining the nature (stochastic vs. deterministic) of the underlying dynamical process that generating the time series.

5.8.4 Kolmogorov-Sinai Entropy

The other commonly used entropy estimation in nonlinear time series analysis is the Kolmogorov-Sinai entropy h_{KS}, which can be obtained from the set of correlation functions $C_m(r)$ (Kolmogorov, 1958; Sinai, 1959). For practical applications, it can be approximated as the limit as the embedding dimension $m \rightarrow \infty$ of the distance (in log-log coordinates) between successive correlation curves $C_m(r)$ and $C_{m+1}(r)$:

$$h_2(m) \cong \lim_{r \to 0} \left(\frac{1}{\Delta t} [\log C_m(r) - \log C_{m+1}(r)] \right) \qquad (5.21)$$

and further

$$h_2 \cong \lim_{m \to \infty}[h_2(m)] \qquad (5.22)$$

One of the main difficulties of extracting the entropies from time series data is mainly because their computation requires far more data than dimensions and Lyapunov exponents, since the limit $m \to \infty$ constitutes the crucial problem (high embeddings are needed). However, as discussed above the h_2 can be approximated using the correlation sum, which is anyway computed for the estimation of the correlation dimension. Furthermore, the upper bound of the Kolmogorov-Sinai entropy can be estimated by the Pesins' identity $h_{KS} = \sum_{i:\lambda_i>0} \lambda_i$ (Eckmann & Ruelle, 1985; Gaspard, 1998).

5.9 Analysis of Stability: Lyapunov Exponents

One of the properties for deterministic chaotic systems is the limited predictability or unpredictability of the future evolution of the system, despite the determinism of the system. The Lyapunov exponents characterize the exponential instability or the average rates of divergence or convergence of nearby trajectories in phase space and therefore measure how predictable or unpredictable the dynamical system is. In other words, they express the loss of information in time and are usually express in units of an inverse of time.

One can estimate as many different Lyapunov exponents for a dynamical system as there are phase space coordinates, i.e. principal axes, which give the average exponential rates of expansion and contraction of the attractor along these axes. Usually in practice, one is interested in the maximal Layapunov exponent that can be used to categorise the type of the motion of the system as presented in Table 5-1.

From the stability analysis we have seen that a positive maximum Lyapunov exponent indicates expansion and exponential divergence of the nearby trajectories. Therefore what distinguishes strange attractors from non-chaotic attractors is the existence of a maximal positive Lyapunov exponent.

TABLE 5-1: POSSIBLE TYPES OF MOTION OF DYNAMICAL SYSTEMS AND THE CORRESPONDING MAXIMAL LYAPUNOV EXPONENTS.

Type of motion	Maximum Lyapunov exponent
Stable fixed point	$\lambda < 0$
Stable limit cycle	$\lambda = 0$
Deterministic chaos	$0 < \lambda < \infty$
Noise (random motion)	$\lambda = \infty$

A formal definition of Lyapunov exponents and their determination for a dynamical system can be described by mathematical equations. Given a continuous dynamical system in d-

dimensional phase space one can monitor the evolution of a set of infinitesimal perturbations of the initial conditions in an attractor that are confined within an d-dimensional sphere (hypersphere), see Figure 5-7. Due to the locally deforming nature of the flow (effects of stretching and folding), this d-sphere will become d-ellipsoid in time. If one orders the principal axes of this sphere (ellipsoid) from the most rapidly to the least rapidly growing, one can compute the average growth (expansion or contraction) rates λ_i ($i=1\ldots d$) of any given principal axis p_i as follows:

$$\lambda_i = \lim_{T \to \infty} \frac{1}{T} \int_0^T dt \frac{d}{dt} \ln\left(\frac{p_i(t)}{p_i(0)}\right) = \lim_{T \to \infty} \frac{1}{T} \ln\left(\frac{p_i(T)}{p_i(0)}\right) \tag{5.23}$$

where $p_i(0)$ is the radius of the principal axis p_i at time $t=0$ (i.e. in the initial hypersphere), and $p_i(T)$ is its radius after some time T.

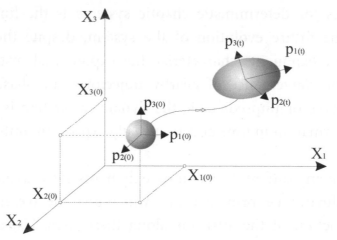

Figure 5-7: A schematic representation of the evolution of a set of initial conditions in the phase space (Velickov, 2004).

When at least one Lyapunov exponent is positive, then the dynamical system is characterized by deterministic chaos, and the initial sphere will evolve to some complex ellipsoid structure reflecting the exponential divergence of nearby trajectories (starting from very similar initial conditions) along at least one direction on the attractor. This sensitivity to small disturbances results in an inability to predict the evolution of the trajectory beyond a certain time horizon, which is approximately the inverse of the divergence rate. However, short-time predictability exists. When no positive Lyapunov exponent exists, then there is no exponential divergence, and thus the long-time predictability of the dynamical system is guaranteed. A set of Lyapunov exponents λ_i is called Lyapunov spectrum. Sano & Sawada (1985) firstly introduced how to calculate Lyapunov spectrum from a time series.

Furthermore, (Kaplan & Yorke, 1979) have conjectured that the dimension of a strange attractor can be approximated from the spectrum of Lyapunov exponents. Such a dimension is called the Kaplan-Yorke (or Lyapunov) dimension and it has been shown that this dimension is close to other dimensions such as the box-counting, information and correlation dimensions for typical strange attractors. The Kaplan-Yorke information dimension is formulated as:

$$D_\lambda = j + \frac{\lambda_1 + ... + \lambda_j}{|\lambda_{j+1}|} \qquad (5.24)$$

where j is the maximum integer such that the sum of the j largest exponents is still non-negative.

5.10 Building Chaotic Model

Once a dynamical system is reconstructed and characterized from the time series of observables, the next general step is to explore the possibilities for constructing models from the data that would realistically model the underlying attractor of the dynamical system. The ultimate goal of constructing such models is prediction, which in the terms of the phase space representation of the dynamics means the extrapolation of the trajectory, thus, modeling the dynamical evolution of the system in time which is yet to be observed. Therefore, in this context, the concept of learning models from data is usually a nonlinear regression estimation of the reconstructed trajectory of the dynamical system from time series data in phase space.

To model nonlinear deterministic dynamics, or a dominant deterministic part of some mixed system, one has to accurately reconstruct the phase space from time series of observables. In this case, an m-dimensional time-delayed embedding based on univariate (scalar) time series of observables is considered; it can be extended latter to vector valued (multivariate) time series embedding. Since the time series data are discretely sampled over time, the underlying dynamics is described by a deterministic model in phase space, which is always a map of the form:

$$Y_{n+1} = f_n(Y_n) \qquad (5.25)$$

where Y_n are delayed vectors (states) $Y_t = \{s_n, s_{n-\tau}, s_{n-2\tau}, ..., s_{n-(m-1)\tau}\}$, formed by the embedding of the time series of observable $\{s_n = x_n + \eta_n\}$ in m-dimensional phase space with an appropriate time delay $\tau = v\Delta t$ (v is an integer time index). In order to prediction the next state of the dynamical system, one needs find the estimator of the regression function $\hat{f}$, and thus, one can estimate the prediction of s_{n+1},

$$\hat{s}_{n+1} = \hat{f}_n(Y_n) \qquad (5.26)$$

After these more general considerations, the next step is to find the proper approximation of the model, expressed through its structure and capacity, and a criterion for the quality of the model which is to be learned from the data in the reconstructed phase space. Generally speaking, there are two possibilities for choosing the structure of the model in order to approximate the mapping function (Casdagli, 1989), namely global and local model approximations. Figure 5-8 illustrates the phase space reconstruction (m=3 and τ=3) and the description of the searching dynamical neighbors and their dynamical trajectories in the past allowing for predicting the future evolution of the dynamical systems in phase space. This example utilizes the real sea level time series data reconstructed in the three-dimensional phase space with time delay τ=3 hours. Three time series samples/points (i.e. 'star' marking) with 3 hours lag in the time domain are represented as a single point in the phase space. Prediction is made by searching the dynamical neighbors (triangle and box markings) of the current point (black circle marking) in phase space and extrapolating the future state by using a local predictive model constructed based on dynamical neighbors.

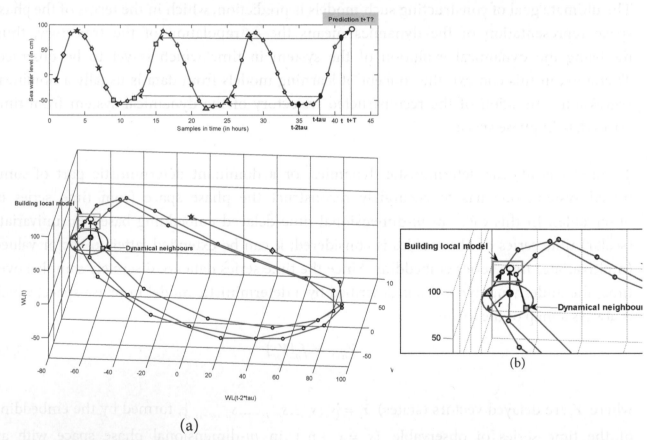

(a)

Figure 5-8: (a) The search of dynamic neighbors and their dynamical evolution in the past predicts the future evolution of the dynamical systems in phase space using local approximation methods. In this example, the real water level time series data at Hoek van Holland tidal station is reconstructed in the three-dimensional phase space (above: in time domain, below: in phase space) (b) The building process of local models approximating the line projections of neighbors into the future states is zoomed in.

The basic idea of the local approximation methods is to use only the states close to present state in phase space in order to make predictions (Farmer & Sidorowich, 1987). Thus, they learn neighborhood relations from the data and map them forward in time. In order to predict the value of the observable x_{n+T}, which is part of the state vector Y_{n+T} where T is some time horizon in the future, based on the state vectors Y_n and past history embedded in the reconstructed phase space, k-nearest neighbors of Y_n are found on the basis of some norm $\|Y_n - Y_{n^*}\|$, with $n^* < n$ (n is a discrete time step). Depending on the number of the neighbors considered and the type of the local mapping chosen, several variations of the local approximation method are possible.

In addition, a multi-step iterative prediction method has been developed and utilized in this work. The multi-step prediction technique consists of making repeated one-step predictions up to the desired horizon. It predicts only one-step ahead using the estimate of the output of the current prediction as the input to the prediction of the next time step until the prediction k-steps ahead is made. The multistep prediction technique demonstrates better prediction performance than the direct prediction method (Box *et al.*, 1994; Kugiumtzis *et al.*, 1998). One of the benefits of using the multi-step prediction is that the dynamical neighbors can be selected iteratively for each one-step prediction. Thus, in most cases, this procedure is able to avoid taking the false neighbors which may produce larger deviations of the neighbor trajectory projections into the future states.

Figure 5-9 illustrates a comparison between direct and multi-step predictions for the surge dynamics in m-dimensional phase space. In this example, we notice beforehand from the observed data that the surge in the next 2 hours would rise up. Suppose trajectory b is the one to be predicted for 2-steps ahead and trajectories a, c and d are the neighbor candidates of trajectory b. The k-nearest neighbors (k-NN) procedure used for finding the neighbors is executed once in direct prediction and h-times (h is the prediction horizon) in multi-step prediction techniques. The trajectory a is a true neighbor and being chosen by both k-NN procedures. On other hand, the trajectory c is a false neighbor which is actually close to trajectory b and selected in the first k-NN procedure, but not in the second k-NN procedure. The trajectory d is the reverse case of trajectory c. Hence, the projection of trajectory b into 2-steps (hours) ahead using direct prediction method produces incorrect prediction (predicting the decreasing surge). This happens due to the inclusion of false neighbor c which subsequently results in building a "false" local model. In contrast, the multi-step prediction is able to predict the increasing surge correctly because the false neighbor c can be avoided (not selected) in the second k-NN procedure.

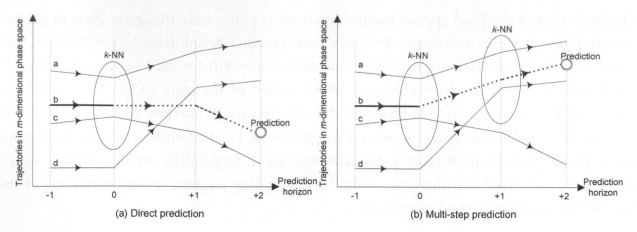

(a) Direct prediction   (b) Multi-step prediction

Figure 5-9: A descriptive comparison between (a) direct prediction and (b) multi-step prediction in m-dimensional phase space. This illustrates that the multi-step prediction can avoid from taking the false neighbor (trajectory c) which may result in wrong projection of the trajectory b to the future states (prediction).

5.11 Recurrence Plots

Recurrence is fundamental feature of many nonlinear dynamical systems. Such a system recurs infinitely many times as close as one wish to its initial state (Poincaré, 1890). The study of recurrences is used to understand the dynamics of nonlinear systems (Marwan *et al.*, 2007). One tool of such purpose is recurrence plot (RP) and its quantifications (Eckmann *et al.*, 1987). RP visualizes the recurrence of states in multi-dimensional phase space into two-dimensional plots. The recurrence of a state at time i at a different time j is pictured within a two-dimensional squared matrix with black and white dots (black dots mark a recurrence, and both axes are time axes). It reveals all the times when the phase space trajectory visits roughly the same area in the phase space. An RP can be mathematically expressed as:

$$R(i, j) = \Theta(\varepsilon - \|\vec{x}(i) - \vec{x}(j)\|), \vec{x}(i) \in \Re^{m}, i, j = 1, \ldots, N \tag{5.27}$$

where N is the number of considered states, ε is a threshold distance, $\|.\|$ a norm (e.g. Euclidean norm) and Θ is the Heaviside step function (see Figure 5-10). An extension of recurrence plots, so-called the recurrence quantification analysis (RQA), is a powerful analytical method developed over the last decade for the study of nonlinear dynamical systems. RQA quantifies the number and duration of recurrences of a dynamical system presented by its phase space trajectory, e.g. recurrence rate, determinism, entropy, laminarity of a dynamical system.

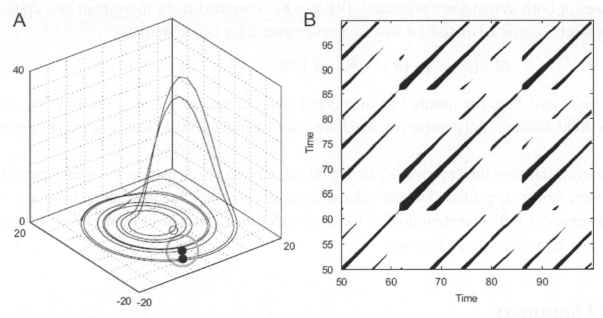

Figure 5-10: An illustration of the phase space trajectory of the Lorenz system (A) and its corresponding recurrence plot (B). A point of the trajectory at j which falls into the neighborhood (circle in (A)) of a given point at i is considered as a recurrence point (black point on the trajectory in (A)). This is marked with a black point in the RP (B) at the location (i, j) (Marwan *et al.*, 2007).

Recently, the recurrence plot has given much intention to researchers in the area of nonlinear dynamics and has been developed into multivariate recurrence plots, e.g. cross recurrence plot (CRP) and joint recurrence plot (JRP) (Marwan *et al.*, 2007). A cross recurrence plot (CRP) is a bivariate extension of the RPs and was introduced to analyze the dependencies between two different systems by comparing their states. CRP shows all those times at which a state in one dynamical system occurs simultaneously in a second dynamical system. In other words, the CRP reveals all the times when the phase space trajectory of the first system visits roughly the same area in the phase space where the phase space trajectory of the second system is. Analogously to the RPs, the cross recurrence matrix is defined by:

$$CR_{i,j}^{\vec{x},\vec{y}}(\varepsilon) = \Theta(\varepsilon - \left\| \vec{x}_i - \vec{y}_j \right\|), i = 1,..., N, j = 1,..., M \qquad (5.28)$$

where N and M is the number of considered states of system x and y, respectively, ε is a threshold distance, $\|.\|$ a norm (e.g. Euclidean norm) and Θ is the Heaviside step function.

For the analysis of two physically different dynamical systems, a joint recurrence plot (JRP) was introduced to analyze the recurrences of their trajectories in their respective pahase spaces separately and look for the times when both of them recur simultaneously. In other words, the JRP is the Hadamard product of the recurrence plot of the first system and the recurrence plot of the second system. By means of this approach, the individual phase spaces of both systems are preserved. JRP can be computed from more than two systems. The joint recurrence matrix for two recurrence matrix for two systems is

$$JR_{i,j}^{\vec{x},\vec{y}}(\varepsilon^{\vec{x}},\varepsilon^{\vec{y}}) = \Theta(\varepsilon^{\vec{x}} - \|\vec{x}_i - \vec{x}_j\|)\Theta(\varepsilon^{\vec{y}} - \|\vec{y}_i - \vec{y}_j\|), i, j = 1,\dots, N \qquad (5.29)$$

where N and M is the number of considered states of system x and y, respectively, ε is a threshold distance, $\|.\|$ a norm (e.g. Euclidean norm) and Θ is the Heaviside step function.

This research uses the recurrence plot technique, aiming at calculating the return period of extreme storm surges based on their dynamics and identifying the interaction between tides and surges as well as meteorological forces (winds and air pressure) by means of CRP and JRP as well as their quantifications.

5.12 Summary

This chapter has shown the fundamental theory of nonlinear dynamics and chaos theory. A simple iterative map can exhibit chaos, similarly in differential equations. A number of tools for the nonlinear analysis of the observed time series have been discussed and can be used for various fields of applications. Many kinds of models representing natural phenomena exhibit chaos. In general sense, it is due to the fact that these natural phenomena behave in chaotic manners. This shows that the methods of nonlinear dynamics and chaos theory are tools for better understanding the complex natural phenomena.

CHAPTER 6: BUILDING PREDICTIVE CHAOTIC MODEL

"Does the flap of a butterfly's wings in Brazil set off a tornado in Texas?"
Edward Lorenz

This chapter explains the construction of predictive chaotic model based on the methods of nonlinear dynamics and chaos theory for predicting storm surges in the North Sea. Several nonlinear time series analysis techniques, such as power spectral density, correlation dimension, mutual information, false nearest neighbors, Lyapunov spectrum are employed to identify the presence of deterministic chaos in the storm surge dynamics and to estimate the proper values of time delays and embedding dimension. Phase space reconstruction and global and local modeling are done for storm surge predictions.

6.1 Introduction

Astronomical tides generally have the large contribution to the ocean water level variations in open oceans and many well-exposed coasts. Traditionally, the analysis of water levels usually employs linear methods that decompose sea levels into tides and other (usually meteorological) components. The amplitudes and phases of the tidal constituents driven by the astronomical motion of the Earth, Moon and Sun (with known periods) can be estimated by using Fourier analysis, response analysis or linear regression methods.

Another component contributing to coastal water level is storm surge. This component can be predicted with an accuracy that depends on the accuracy of the meteorological predictions. An appropriate numerical weather model can predict the motion of atmospheric depression with a satisfactory accuracy in a range of several days. The wind and air surface pressure fields predicted by this model can be utilized as some driving forces of the sea motion in a shallow water model allowing for storm surge predictions.

A lot of research has been conducted on understanding and modeling oceanic water level have been made for more than a century (see also Chapters 2 and 3). Korteweg & de Vries

(1895) characterized the weakly nonlinear shallow water waves by the Korteweg-de Vries (KdV) equation which is an exact solvable partial differential equation. Zabusky & Kruskal (1965) found that the solitary wave solutions from KdV equation can be obtained in the continuum limit of the Fermi-Pasta-Ulam Experiment (Fermi *et al.*, 1955) and have similar behavior to the superposition principle, despite the fact that the waves themselves were highly nonlinear. However, the water level dynamics in coastal and estuarial swallow-water areas, such as the Dutch coast, may differ significantly from the astronomical estimated constituents (superposition principle) – due to the nonlinear effects that include meteorological forcing, tidal- current interactions, tidal deformations due to the complex topography and river discharges (Prandle & Wolf, 1978; Otto *et al.*, 1990; Horsburgh & Wilson, 2007). A number of models have been built (some of them are characterized in Chapter 3) but a lot of research in this area is still in progress.

The ocean water level variations due to various determinants and their complex interactions show long-term persistence leading to the correlated extreme events (Alexandersson *et al.*, 1998; Butler *et al.*, 2007). Complexity of the described phenomena prompts for adequate methods to describe them, and one of them is chaos theory (Dijkstra, 2005). The most direct link between the concept of deterministic chaos and the real world is the analysis of data (time series) from real systems in terms of the theory of nonlinear dynamics (Tsonis, 1992; Abarbanel, 1996; Donner & Barbosa, 2008). Note that this approach is, in fact, data-driven, since it is mainly based on the analysis of the observation data. The initial nonlinear analyzes of the ocean water levels at the Florida coast have been conducted by Frison *et al.* (1999). The early experiments of the use of predictive chaotic model (CM) for storm surge predictions were done by (Solomatine *et al.*, 2000) and (Walton, 2005) using univariate local models. Velickov (2004) extended the method using multivariate predictive chaotic models and showed that it has reliable and accurate short-term predictions. Nonlinear and exploratory analysis, predictability, entropy, complexity of storm surge dynamics in the North Sea have been also explored in the UNESCO-IHE Master studies by Pupo (2000), Hasan (2001), and subsequent papers by Solomatine *et al.* (2001) and Velickov *et al.* (2003).

This chapter presents the use and implementation of the methods described in the previous chapter – nonlinear dynamics and chaos theory – for predicting storm surges. If compared to earlier works (Solomatine *et al.*, 2000; Solomatine *et al.*, 2001; Velickov *et al.*, 2003; Velickov, 2004), we advanced the procedure of building predictive chaotic model by incorporating several new features: using Cao's method (Cao, 1997) for better estimation on dynamical invariants; implementing multi-step iterative predictions, applying the Euclidean distance threshold to avoid inclusion of the false dynamical neighbors, adding water level variable into multivariate predictive chaotic models, finding the proper number of neighbors using performance-based technique for stormy and non-stormy conditions, and optimizing the predictive chaotic model parameters (time delay and embedding dimension). Furthermore, we compare the prediction performances of the proposed predictive chaotic model with other models, including artificial neural network (ANN) models. For chaotic analysis of non-linear time series we used TISEAN software (Hegger *et al.*, 1999; Kantz & Schreiber, 2004), other statistical analysis software and some specially written MATLAB scripts, and for building predictive models – dedicated software components developed in MATLAB.

6.2 Power Spectral Density: Periodicity and Stochasticity

Power spectral density function (PSD) shows the strength of the variations (energy) as a function of frequency. It describes how the power (or variance) of a time series is distributed with frequency (Emery & Thomson, 2001; Broersen, 2006). Intuitively, the spectral density captures the frequency content of a stochastic process or time series (e.g. wind wave data) and helps identify periodicities. The unit of PSD is energy per frequency (width) and you can obtain energy within a specific frequency range by integrating PSD within that frequency range. Computation of PSD is done directly by the method called Fast Fourier Transform (FFT) or computing autocorrelation function and then transforming it. Figure 6-1 shows the periodogram power spectral densities of the water level (left) and surge (right) time series data at Hoek van Holland tidal station. Both power spectral densities display a broadband spectral distribution with some sub-harmonic components observed.

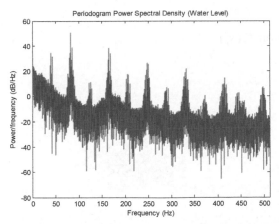 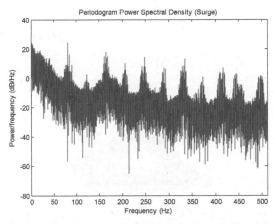

Figure 6-1: Power spectral density of the water level (left) and surge (right) time series data at Hoek van Holland tidal station. Both periodograms display a broadband spectral distribution with some sub-harmonic components observed.

6.3 Phase Space Reconstruction: Finding Time Delay

As described in Chapter 5, the most important phase space reconstruction technique is the method of delays, which is associated with the Taken's embedding theorem (Packard *et al.*,

1980; Takens, 1981). Vectors in a new space, the embedding space are formed from time delayed values of the scalar measurements. According to Taken's theorem, the dynamics of a time series $\{x_1, x_2, ..., x_N\}$ are fully captured or 'embedded' in the m-dimensional phase space defined by the delay vectors $Y_t = \{x_t, x_{t-\tau}, x_{t-2\tau}, ..., x_{t-(m-1)\tau}\}$ where τ is the delay time.

We have tested several values of delay and plot the phase portrait for embedding dimension $m=3$ (this is the maximum dimension that is possible to visualize). For illustration, Figure 6-2 depicts the results water level and surge time series.

In real applications, the delay time τ needs to be appropriately chosen in order to fully capture the structure of the attractor. If τ is too small then the delay vectors are not independent, such that all points are accumulated around the bisector of the embedding space, resulting in loss of characteristics on the attractor structure. If τ is very large, the different coordinates (delay vectors) may be almost dynamically uncorrelated. The straightforward choice of τ is usually made with the help of the zero-crossing autocorrelation function. However, in terms of nonlinear methods, the choice of τ corresponding at the first minimum of the time delayed mutual information demonstrates good performance in reconstructing the system dynamics from time observables. This mutual information is based on the Shanon's entropy (Fraser & Swinney, 1986) and can be computed as follows: Given a time series of observable s, one can calculate the transitional probabilities $P_s(s_i)$ that a measurement s yields s_i. The information entropy is thus defined as:

$$H(s) = -\sum_{i=1}^{N} P_s(s_i) \log P_s(s_i)$$

(6.1)

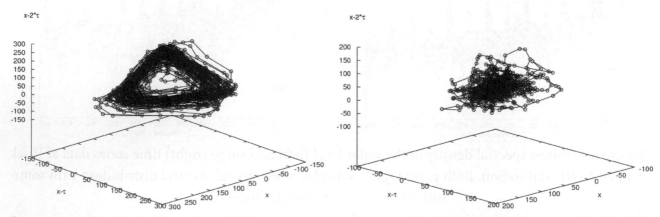

Figure 6-2: The three-dimensional phase space reconstruction for the 1000 data points of the hourly water level (left, $\tau=4$, $m=3$) and surge (right, $\tau=10$, $m=3$) time series data at Hoek van Holland tidal station.

Figure 6-3 shows the autocorrelation and mutual information the water level and surge time series data at Hoek of Holland (1990-1996). The first minimum of the mutual information which characterizes the nonlinear relationships between time-lag variables is found to be a better criterion than the zero crossing autocorrelation (only measures linear dependency) for the choice of optimal time delay in the phase space reconstruction of time series data. The first minimum values of mutual information suggesting for the optimal time delay τ are 4 and 10 hours for water level and surge time series data, respectively.

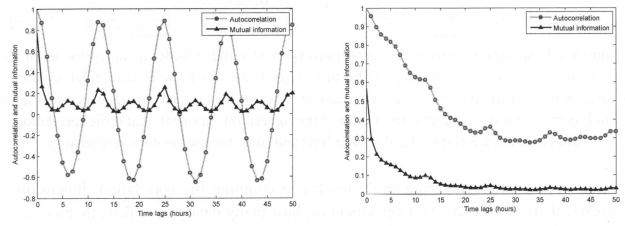

Figure 6-3: The autocorrelation function (dotted line with circles) and the mutual information (solid line with triangles) as a function of time lags for the hourly water level (left) and surge time series at Hoek van Holland tidal station. The nonlinear correlation illustrated by mutual information indicates the optimal time delays are 4 and 10 hours for water level and surge time series data, respectively.

Note that the described methods is not the only one: in Chapter 8 we employ more "pragmatic" optimization methods to find delay τ and embedding dimension m, such that they would maximize the performance of the predictive model.

6.4 Correlation Dimension

Figure 6-4 shows that the correlation exponent increases with an increase of the embedded dimension up to a certain value and further saturates. The saturation values of the correlation exponents/dimensions using the optimal time delays of 4 and 10 hours are 6.5 and 8.5 for the water level and surge time series data, respectively. This indicates the presence of an attractor in the water level and surge dynamics. Taking into account the previous discussion about the estimation of the embedding dimension m, if one uses the Taken's embedding theorem, the embedding dimensions ($m=2d_c+1$) of the manifold which contains the attractor are 15 and 18 for the water level and surge dynamics. Kennel *et al.* (1992) suggests the minimum embedding dimension of $m \geq d_c$. This specifies that the embedding dimensions of 6(7) and 8(9) are enough to unfold the water level and surge attractors, respectively. These results, however, need to be verified by other embedding dimension estimators as described in the following sections.

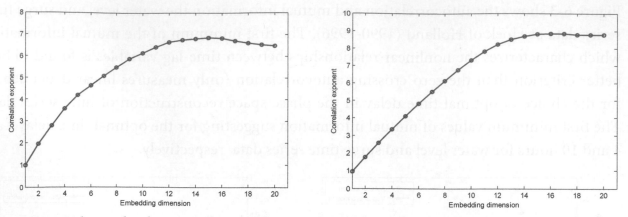

Figure 6-4: Relationship between the correlation exponent ν and embedding dimension m for the hourly water level (left) and surge (right) time series data at Hoek van Holland tidal station. Correlation exponent increases with an increase of the embedded dimension up to a certain value and further saturates. The saturation value of the correlation exponent, that is the correlation dimension, is 6.5 and 8.5 for the water level and surge time series data, respectively.

A large size of data set is commonly needed to compute the correlation dimension d_c. However, there is no consistent agreement on how many data can sufficiently provide the accurate estimation of the correlation dimension. Some authors like (Smith, 1988), (Theiler, 1990) and (Ruelle, 1990) suggest differently on the minimum size of data set required for estimating correlation dimension. For correlation dimension 8.5, the size of data set used here is sufficient to estimate the correlation dimension (Ruelle, 1990). The size of 54768 data points of the hourly surge time series from 1 January 1990 till 31 March 1996 is larger than the minimum data set size $10^{dc/2}$ suggested by Ruelle (1990) which is about 17783 data points. Please also note that the data set in this work was obtained from the real observations representing the physical processes in nature, and not merely on the basis of a uniform-random model. Nonetheless, we consent that the larger size of data sets might be needed for better estimation of correlation dimension.

6.5 False Nearest Neighbors

The false nearest neighbor (FNN) method can determine the minimal sufficient embedding dimension m (Kennel *et al.*, 1992). The false neighbors are the points projected into neighborhoods of other points to which they do not belong as neighbors in higher dimensions. Figure 6-5 shows that the percentage of the FNN drops to about 1% with the embedding dimensions of $m=6$ and 8 for water level and surge time series and remains unchanged for a further increase in the embedding dimension. This result is consistent with the estimation using correlation dimension based on the rule of (Kennel *et al.*, 1992) suggesting that the minimum embedding dimension is $m>d_c$.

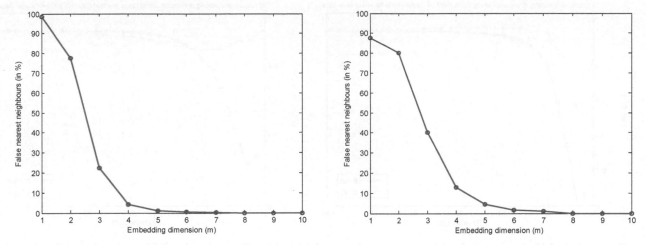

Figure 6-5: The percentage of the false nearest neighbors (FNN) as a function of the embedding dimension for the water level (left) and surge (right) hourly time series data at Hoek van Holland tidal station. The FNN suggests the optimal embedding dimensions for water level and surge time series data are 6 and 8, respectively.

6.6 Cao's Embedding Dimension

The FNN algorithm has a drawback associated with the subjective choice of the threshold in order to ensure a correct distinction between low-dimensional chaotic data and noise. To avoid this issue, Cao's method or the averaged false neighbors (AFN) method (Cao, 1997) was utilized to find the proper embedding dimension for water level and surge dynamics at HvH. Cao's approach is based on the estimation of two parameters $E1$ and $E^\star$ which are basically derived from quantities that are defined by the FNN method. These parameters are computed for different increased values of the embedding dimension m. Then the global behaviors of $E1$ and $E^\star$ as functions of dimension m are respectively used to estimate the minimum embedding dimension and to determine the nature stochastic vs deterministic of the underlying dynamical process generating the time series. This method has many advantages: it does not need too long time series, it is computationally efficient, and some of its features are not very sensitive to noise. Moreover, it is not based on any arbitrary choice of a threshold. Figure 6-6 shows the saturated lines of $E1(m)$ can be obtained starting from dimensions m of 6 and 8 for water level and surge time series data at HvH. Number of neighbor (k) was set to 1. There is no existence of the straight lines of $E^\star(d)$ indicating that water level and surge dynamics are not purely driven by random behaviors.

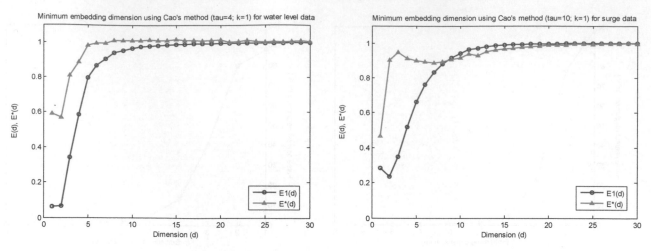

Figure 6-6: Minimum embedding dimension estimated by Cao's method for water level (left) and surge (right) data at Hoek van Holland tidal station. The Cao's method also suggest the embedding dimensions of 6 and 8 for water level and surge time series data, respectively.

6.7 Space-Time Separation

Another technique for identifying temporal correlations inside the time series or determining a reasonable time delay is space-time separation plot (Provanzale *et al.*, 1992). This method integrates along parallels to the diagonal and thus only shows relative times. One usually draws lines of constant probability per time unit of a point to be a neighbor of the current point, when its time distance is δt. In other words, it shows how large the temporal distance between points should be so that we can assume that they form independent samples according to the invariant measure. This plot is also useful to detect stationary and give a warning when the data points are too few. Figure 6-7 depicts the space-time separation plot of water level (left) and surge (right) using $\tau=4$, $m=6$ and $\tau=10$, $m=8$, respectively. It is clearly seen that the M2 semi-diurnal tidal constituent with period 12.42 hours plays an important role in water level dynamics.

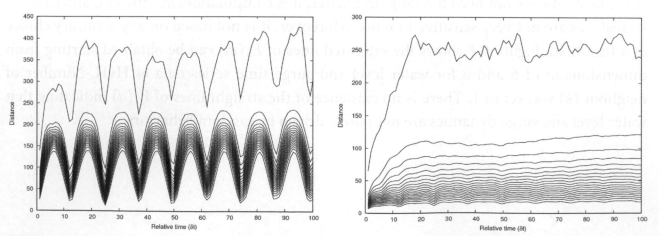

Figure 6-7: Space-time separation plots for water level (left, $\tau=4$, $m=6$) and surge (right, $\tau=10$, $m=8$) time series data at Hoek van Holland tidal station.

6.8 Lyapunov Exponents

The Lyapunov spectrum (Sano & Sawada, 1985) estimated from the water level and surge time series at Hoek van Holland tidal station is presented in Figure 6-8. The largest Lyapunov exponent is estimated as λ_1=0.08 for both water level and surge time series which indicates a loss of information of 0.08 bits/hour during the dynamical evolution of the system, and thus loss of predictive capabilities. The Lyapunov spectrum contains a large negative exponent λ_6=-0.4 and -0.75 for water level and surge time series, respectively, which indicates presence of strong dissipation mechanisms in the dynamics of the system. The presence of positive Lyapunov exponents and the fact that sums of Lyapunov exponents are negative (-1.48 for both time series), provide strong evidence that water level and surge dynamical systems in the North Sea are driven by deterministic chaos. Furthermore, the Kaplan-Yorke information dimensions for water level and surge time series were estimated to be 2.77 and 4.1, respectively. The existence of a fractal Kaplan-Yorke information dimension indicates deterministic chaos in the dynamical system.

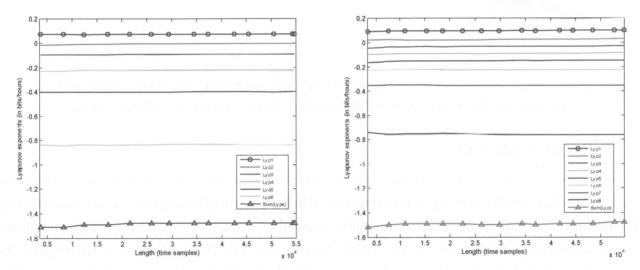

Figure 6-8: The Lyapunov spectrum for the water level (left, m=6) and surge (right, m=8) time series data at Hoek van Holland tidal station. The spectrum are consistent showing the largest Lyapunov exponents (lines with circles) are positive and the sum of global Lyapunov exponents (lines with triangle) are negative, for both time series.

6.9 Poincaré Sections

A very convenient way to delineate the dynamical system is given by Poincaré sections. A Poincaré section is a slice obtained from the intersections of trajectories in m-dimensional attractor with an (m-1)-dimensional surface in the phase space. The usefulness of the Poncaré section lies in the reduction of order of the dynamical system and it bridges the gap between continuous and discrete-time systems. If one deals with periodic evolution of period n, then this sequence consists of n dots repeating in the same order. If the evolution of the system is quasi-periodic the sequence of points defines a closed limit cycle. Finally, if the evolution is deterministic chaos, then the Poincaré section is a collection of points that show an interesting pattern, often revealing the fractal nature of the underlying attractor. Figure 6-9 depicts the Poincaré section of the first 10000 data points hourly water level and surge at Hoek van Holland tidal station with $m=6$, $\tau=4$ and $m=8$, $\tau=10$, respectively, as estimated by nonlinear chaotic analysis.

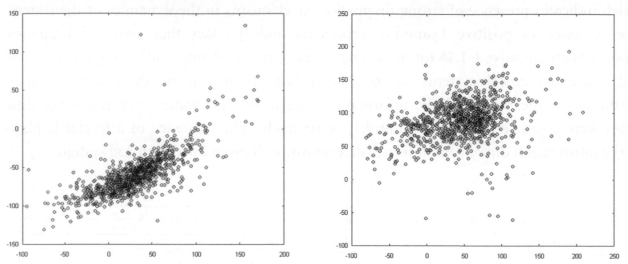

Figure 6-9: Poincaré sections of hourly water level (left, $m=6$, $\tau=4$) and surge (right, $m=8$, $\tau=10$) time series data using the first 10000 data points at Hoek van Holland station.

6.10 Recurrence Plot

One of powerful tools for visualizing the recurrences of a dynamical system is a recurrence plot (RP). It exploits the dynamical system's behavior in phase space (such as detecting transition, dynamical invariant, unstable periodic orbit) by means of 2-dimensional visualization/plot (Eckmann *et al.*, 1987; Marwan *et al.*, 2007). Figure 6-10 depicts the recurrence plots for water levels and surges at Hoek van Holland location. It is seen similar extreme storms in the dynamics occurred as indicated by white bands.

Further development of this technique is to apply the recurrence plot for identifying the return period of extreme storm surges. A new technique combining recurrence plot and peak over threshold (POT) (Castillo *et al.*, 2005) is proposed. The main procedure of this technique consists of (see Figure 6-11):

(a) Set a threshold for the Euclidean distance in unthresholded RP ($\varepsilon>t$);

(b) Cluster extreme values from RP with size depending on the values of embedding dimension and time delay;

(c) Find the relevant index of extreme values of each cluster in real data;

(d) Locate the maximum values in each cluster;

(e) Feed these maximum values into POT analysis.

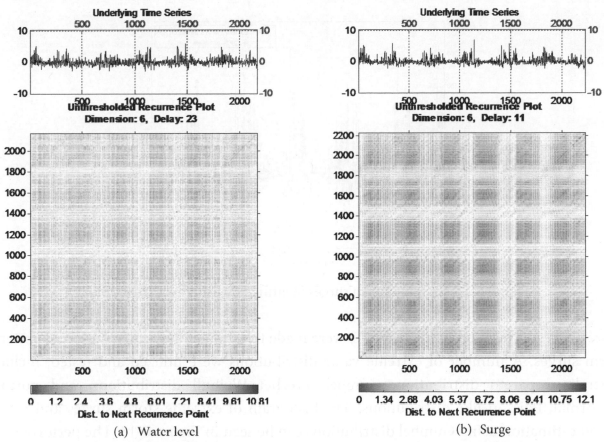

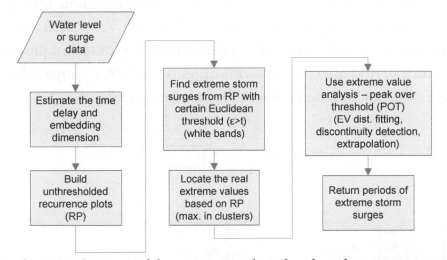

(a) Water level

(b) Surge

Figure 6-10: Recurrence plots of water levels and surges at Hoek van Holland location.

Figure 6-11: A schematic diagram of the main procedure for identifying return period of extreme storm surges using recurrence plot (RP) and peak over threshold (POT).

For experiments, the water level and surge data at Hoek van Holand location was used. The recurrence plot was computed with parameters ($\varepsilon > 10.5$; max clustering; m=6; $\tau=11$). Figure 6-12 shows the identification of extreme storm surges using recurrence plot.

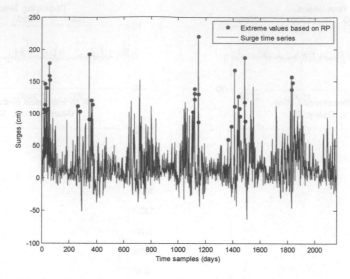

Figure 6-12: Extreme storm surges identified by using recurrence plot.

More processes (as seen in Figure 6-11) were made to estimate the return period of extreme storm surges. A number of extreme value distributions were utilized and tested, include: generalized pareto distribution; Gumbel, Frechet Weibull distributions; and normal, lognormal, Poisson-like distributions. The histogram of extreme storm surges and return period estimation using Gumbel distributions can be seen in Figure 6-13. The performances of methods of recurrence plot and extreme value analysis in estimating the return periods of water level and surge are listed in Table 6-1. The best distribution fitting is achieved by using Poisson-like distribution for the method of extreme value statistics (EVS) whereas Gumbel distribution is for recurrence plot. The recurrence plot techniques can improve the estimation of return period of extreme storm surges (CoE=0.99) compared with the method of extreme value statistics; however it is not the case for water level.

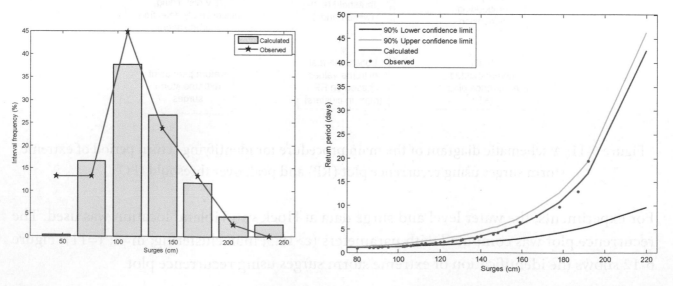

Figure 6-13: (a) Histogram of extreme storm surges and (b) return period estimation using Gembel distributions.

TABLE 6-1: PERFORMANCE OF METHODS OF RECURRENCE PLOT AND EXTREME VALUE STATISTICS FOR ESTIMATING RETURN PERIODS OF EXTREME WATER LEVELS AND SURGES.

Met.	Data	Dim., delay m,τ	Thres. t,ε	N	EV dist. fitting	Return Periods (days, cm)					MAE of CDF	CoE
						7	30	60	100	365		
EVS	Sur	NaN	80	107	Poisson-like	166	212	234	249	289	0.017	0.94
	WL	NaN	150	227	Poisson-like	238	280	300	314	350	0.015	0.98
RP	Sur	11,6	10.5	38	Gumbel	134	176	196	210	247	0.019	0.99
	WL	11,23	9.5	45	Gumbel	202	241	260	274	309	0.036	0.94

6.11 Predictive Chaotic Model: Global and Local Modeling

Section 5.10 has described the process of building of predictive chaotic model based on the identified and reconstructed dynamical system. Two possibilities of local and global modeling can be utilized. With respect to data-driven modeling techniques, the global and local models can be built from any type of computational intelligence (CI) predictive models, like ANN or radial-basis functions (Haykin, 1999). The main different concepts on global and local modeling in the view of machine learning perspective (Mitchell, 1997) is that global modeling tends to be eager learning whereas local modeling is lazy or instance-based learning (Aha *et al.*, 1991). Hence, in global modeling, the model or abstraction is firstly constructed based on the available data (in this case it is the reconstructed phase space matrix) before prediction. In contrast, the local modeling keeps the presented training data (the reconstructed phase space) and waits until prediction is requested and the query instance is the last point in the reconstructed phase space. The *k*-nearest neighbor method are a common approach in the instance-based learning to approximating real-valued or discrete-valued target functions. The neighbors found by *k*-NN algorithm are then used for building the local models which can be any types of predictive models. In this work, the zeroth (constant), linear, quadratic and 3rd-order polynomial models are used as local models for predicting storm surge dynamics at Hoek van Holland station.

Figure 6-14 depicts the chaotic model predictions for predicting storm surge at Hoek van Holland station during stormy period, where the dynamical neighbors found by *k*-NN algorithm are projected into 3-steps ahead (right). The number of neighbors is different for each step of prediction depending on the presence of similar behavior of storm surges in the past (dynamical neighbors). Subsequently, these neighbors are used for constructing the local models to approximate the future surge condition.

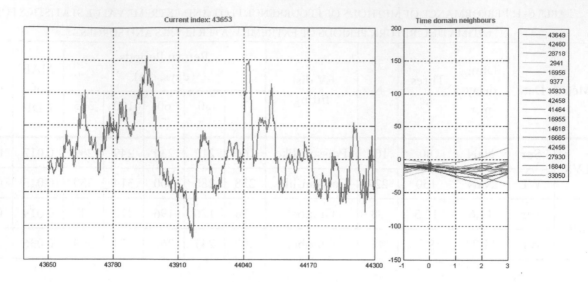

Figure 6-14: Chaotic model predictions with dynamical neighbors projected into 3-steps ahead.

6.12 Model Setup

6.12.1 Univariate predictive chaotic model

The nonlinear chaos analysis (correlation dimension, FNN and Cao's method) of water level and surge time series recommends the appropriate values of time delay and embedding dimension are τ=4, m=6 for water level and τ=10, m=8 for surge. Yet, these estimations do not consider the effect of the time delay and embedding dimension selections to the chaotic model prediction performance. Therefore, the exhaustive search optimization was used for finding the optimal values of the embedding dimension (m), the time delay (τ) and the number of neighbors (k). In view of the fact that the exhaustive search optimization for the predictive chaotic model requires very intensive computation, we utilized sensitivity analysis to search for the appropriate number of neighbors (k) for each non-storm and storm conditions of water level and surge. After these values are obtained, we execute the exhaustive search for finding the optimal values of time delay and embedding dimension.

The sensitivity analysis was done by setting up the predictive chaotic model parameters for the water level (with τ=4 and m=6) and surge (with τ=10 and m=6) and the number of neighbors (k) run from 1 to 2000. We use 3rd-order polynomial local models which are built based on the dynamical neighbors. In the internal procedure for finding the neighbors, some neighbors which have distance further than twice (adjustable by the user) of the nearest neighbors obtained should be cut out for removing the possible false neighbors. The procedure is as follows:

(1) Define the number of neigbhours (k),

(2) Find the k-nearest neigbhours in the phase space,

(3) Cut the discovered nearest neighbors that have distance more than twice of the nearest neighbors.

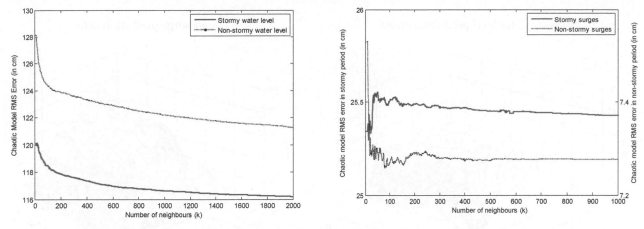

Figure 6-15: The six-hours prediction error of the predictive chaotic models as a function of the number of neighbors (k) for non-stormy and stormy water level (left, τ=4, m=6) and surge (right, τ=10, m=6), respectively, at Hoek van Holland tidal station.

Figure 6-15 depicts the six-hours prediction RMS errors of the predictive chaotic models as a function of the number of neighbors (k) for non-stormy and stormy water level and surge time series data. It is clearly shown that the suitable number of neighbors for predicting surges during storm condition is small (13 neighbors) and it should be smaller than the one (80 neighbors) during non-storm condition for surges. One of the reasons is that less true dynamical neighbors (similar surge behavior in the past) can be found especially during extreme storms. If we take more neighbors, the model performance will be worse due to the inclusion of false neighbors in constructing local models. Consequently, the whole predictive chaotic model performance will decrease. For the water level, however, the model error decreases as the number of neighbor increases. This indicates that the use of global approximation model for predicting water level would be better than the use of local model. The reason is that the large part of water level components comes from tides which are quasi periodic and well predicted by harmonic analysis.

The exhaustive search optimization was done with the following settings: time delay range=[1~24], embedding dimension range=[2~30], 3rd order polynomial local model and the number of neighbors k=13 and k=80 for surges during stormy and non-storm conditions, respectively, and k=300 for water level. The prediction horizons are 1, 3, 6, 10 and 12 hours. Each prediction horizon can have different values of time delay and embedding dimension. The optimization outcome is the most accurate predictive chaotic model which has the lowest RMS error on cross validation data set. The cross validation data sets have small size of 400 data points: time indices of 35500-35900 for storm condition and 38200-38600 for non-storm condition. This small size of cross validation data sets was employed with considerations of the necessity of intensive computation for exhaustive search.

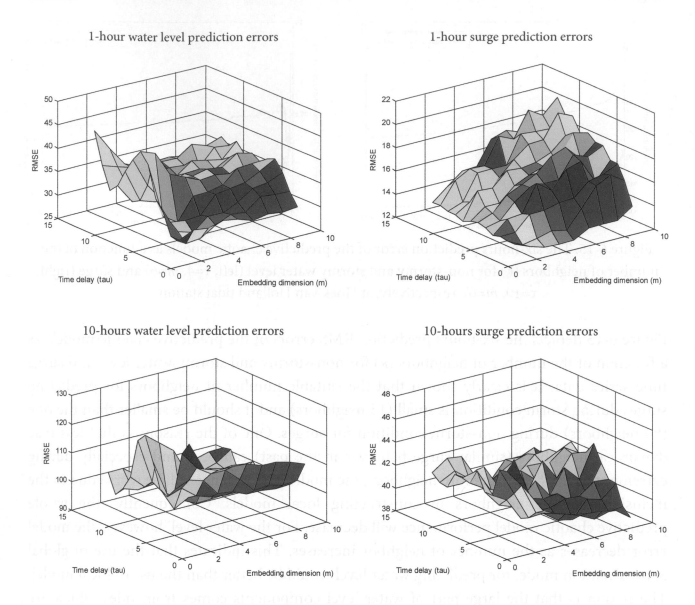

Figure 6-16: The 3D surface of the univariate predictive chaotic model RMS errors for 1 and 10 hours prediction horizons during stormy period (time index: 35500-35900) as a function of time delay and embedding dimension for water level (left, τ=4, m=6) and surge (right, τ=10, m=8) time series data at Hoek van Holland tidal station.

Figure 6-16 illustrates the 3D surface of the univariate predictive chaotic model RMS errors for 1 and 10 hours prediction horizons during stormy period (time index: 35500-35900) as a function of time delay and embedding dimension for water level (left, τ=4, m=6) and surge (right, τ=10, m=8) time series data at Hoek van Holland tidal station. The 3D surfaces are unlike for each prediction horizon and storm/non-storm condition. This denotes that the choice of time delay and embedding dimension for phase space reconstruction should consider these variables. For example, the optimal time delay and embedding dimension for univariate predictive chaotic model for predicting 6 hours ahead surges during storm condition was obtained:

$$\mathbf{Y}_{t+6} = \{s_t^{hvh}, s_{t-3}^{hvh}, s_{t-6}^{hvh}, s_{t-9}^{hvh}, s_{t-12}^{hvh}, s_{t-15}^{hvh}\} \tag{6.2}$$

The complete optimal univariate predictive chaotic model structures are listed in Table 6-2.

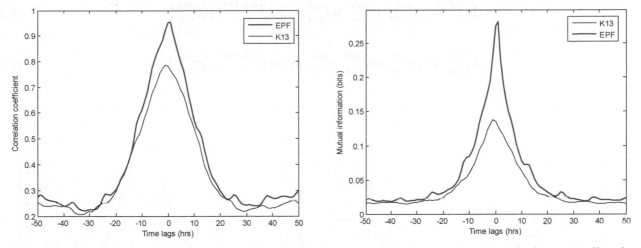

Figure 6-17: The cross correlation and mutual information between surges at Hoek van Holland and neighboring stations (EPF and K13). Both techniques show that the EPF surges precedes surges at HvH about 1 hour and the K13 surges has less relationship with HvH surges and the HvH surges would reach to K13 around 1-1.5 hours later.

6.12.2 Multivariate predictive chaotic model

Multivariate predictive chaotic models incorporating information on water level and surge at Hoek van Holland and neighboring stations (EPF and K13), air pressure (difference) and wind components were employed with the main objective to improve the prediction accuracy for longer prediction horizons. The relationship between water level and surge at Hoek van Holland and EPF/K13 are measured by cross correlation and mutual information as shown in Figure 6-17. Both methods specify that the EPF surge precedes the surge at HvH about 1 hour and the K13 surge has less relationship with HvH surge and the HvH surge would reach to K13 around 1-1.5 hours later. Thus, we include the information from EPF as inputs of predictive chaotic model.

The other variables which require more analysis are wind speed and direction. Cross correlation and mutual information were applied for acquiring the principal wind component which has largest influence to the surge at Hoek van Holland. The various wind directions from 0 to 180 degrees from North were investigated. The strongest influence of the winds on the surge (correlation coefficient=-0.65) is generated by wind component 120 degree from North (Figure 6-18). Likewise, it is indicated by mutual information.

The multivariate phase space reconstruction of the surge dynamics using hourly time series data was solved technically using the multivariate embedding. The exhaustive search optimization was also utilized to all possible combinations of time delay and embedding dimension for each observable. The phase space structures for water level and surge are as follows:

$$Y_{wl}^{hvh} = \{s_{\tau,m}^{hvh}, wl_{\tau,m}^{hvh}, wl_{\tau,m}^{epf}, wind_{\tau,m}^{hvh,120}, dp_{\tau,m}^{hvh}\} \tag{6.3}$$

$$Y_{surge}^{hvh} = \{s_{\tau,m}^{hvh}, wl_{\tau,m}^{hvh}, s_{\tau,m}^{epf}, wind_{\tau,m}^{hvh,120}, dp_{\tau,m}^{hvh}\} \tag{6.4}$$

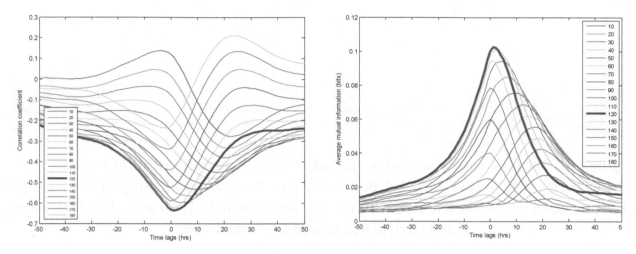

Figure 6-18: The cross correlation and mutual information between wind components and surge at Hoek van Holland with various wind direction (0-180 degrees from North). The strongest influence of the winds on the surge (correlation coefficient=-0.65) is generated by wind component 120 degree from North (left). Similarly, it is indicated by mutual information (right).

The difference between these two structures is located at the inclusion of the information from neighboring station. In this case, we utilized the water level and surge from EPF tidal station. The water level phase space uses the water level from EPF, while the surge phase space requires the surge from EPF. The next task is to optimize the values of the τ and m of each variable for using exhaustive search. The exhaustive search optimization was done with the following settings. The time delay and embedding dimension for $s_{\tau,m}^{hvh}$ and $wl_{\tau,m}^{hvh}$ are fixed with the values as obtained from nonlinear chaos analysis, $s_{\tau,m}^{hvh}$ has $\tau=10$ and $m=8$, while $wl_{\tau,m}^{hvh}$ has $\tau=4$ and $m=6$. The time delay and embedding dimension for other variables ($wl_{\tau,m}^{epf}, s_{\tau,m}^{epf}, wind_{\tau,m}^{hvh,120}, dp_{\tau,m}^{hvh}$) are: τ range=[1~5] and m range=[2~5]. We employed the 3rd order polynomial local model and the number of neighbors of $k=13$ and $k=80$ for surges during stormy and non-storm conditions, respectively, and $k=300$ for water level. The prediction horizons are 1, 3, 6, 10 and 12 hours. The optimization outcome is the most accurate predictive chaotic model which has the lowest RMS error on cross validation data set.

The optimal multivariate predictive chaotic model structure can be seen in Table 6-4 for storm surge and Table 6-5 for water level. For example, the optimal multivariate phase space reconstruction for predicting surges with 3 hours prediction horizon during storm condition was obtained:

$$\mathbf{Y}_{t+3} = \{s_t^{hvh}, s_{t-10}^{hvh}, ..., s_{t-70}^{hvh}, wl_t^{hvh}, wl_{t-4}^{hvh}, ...wl_{t-18}^{hvh}, s_t^{epf}, s_{t-3}^{epf}, ..., s_{t-9}^{epf}, ...$$
$$..., wind_t^{120^o}, wind_{t-2}^{120^o}, dp_t, dp_{t-1}, ..., dp_{t-3}\} \tag{6.5}$$

For predicting surges during storm condition, the optimization results indicate that the most appropriate time delay for $s_{\tau,m}^{epf}$ is 1 hour for all prediction horizons. This coincides with the analysis as depicted in Figure 6-17 viewing that the EPF surges precedes surges at HvH about 1 hour.

6.12.3 Global model: Neural networks

Backpropagation multi-layer perceptrons (MLPs) with Levenberg-Marquardt training rule (Haykin, 1999) was utilized and trained using the same input structure as the predictive chaotic model inputs. The number of hidden neurons of ANN was selected using the exhaustive search with range of [1~10]. The optimal MLPs structures are listed in Table 6-3 for univariate NN, and Table 6-4 and Table 6-5 for multivariate NN.

6.13 Model Results and Discussion

Table 6-2 and Table 6-3 summarize the univariate predictive chaotic model and neural network model prediction performances. The optimal parameters for predictive chaotic models (m, τ, k) and neural network models (the number of hidden neurons) were obtained based on the model performance (RMS error) on cross validation data sets for non-stormy and stormy periods and we tested these models on testing data set with various optimal values of m, τ, k and number of hidden neurons based on the prediction horizons and storm and non-storm condition. These comparison results (up to 12 hours prediction) showed, in general, the predictive chaotic model and neural network model have similar prediction accuracy either during stormy or non-stormy periods. However, the predictive chaotic model is able to reach the extreme surges better than the neural network model in predicting surges. This is depicted in Figure 6-19. The predictive chaotic model errors are more dampened and stable (including during the surge peaks) than the neural network errors.

TABLE 6-2: PERFORMANCE OF UNIVARIATE PREDICTIVE CHAOTIC MODEL WITH PARAMETERS OPTIMIZED BY EXHAUSTIVE SEARCH.

Predictive chaotic models		Prediction horizons				
		1 hrs	3 hrs	6 hrs	10 hrs	12 hrs
Surges						
Stormy (k=13)	τ	1	1	3	1	2
	m	5	7	6	10	6
	RMSE(CV)	12.35	12.91	24.38	38.51	45.15
	RMSE(Test)	11.11	**11.94**	**21.69**	30.44	34.42
Non-stormy (k=80)	τ	1	1	1	4	12
	m	9	5	7	6	9
	RMSE(CV)	5.09	5.35	7.56	9.04	9.94
	RMSE(Test)	5.87	**6.00**	8.46	10.81	11.86
Water levels						
Stormy (k=300)	τ	2	6	6	6	11
	m	3	3	4	5	3
	RMSE(CV)	27.25	37.54	90.61	97.45	74.59
	RMSE(Test)	26.66	39.20	91.30	96.37	66.72
Non-stormy (k=300)	τ	3	6	8	8	11
	m	4	3	9	4	3
	RMSE(CV)	23.93	36.17	89.20	91.00	60.16
	RMSE(Test)	23.39	36.10	86.85	90.32	59.15

However, the predictive chaotic models performed worse than the neural network models for predicting water levels. In fact, their predictions are generally not quite good. The main reasons of not accurate predictions in predictive chaotic model are that the water level contains sharp oscillations which are difficult for the 3rd order polynomial local model to approximate with. Moreover, the building of this local model may include many false neighbors due to the fact that the trajectories are very close each other and the nearest neighbors found most possibly have different or reverse directions of trajectories. The other cause is due to the inherent issues with nonlinear discrete time series which we can build the phase space with integer (not fractal) values of time delay and embedding dimension. As we obtained from the space-time separation plot, the period of water level is 12.42 hours (non integer) which is not similar with the nonlinear analysis (mutual information) outcome suggesting time delay of 4 hours (integer). This issue results in the amplification of shift errors as the prediction horizon increases. It is clearly seen in Figure 6-19 where the predictive chaotic model errors oscillate sharply due to the existence of phase errors. The possible solutions for these are: to use smaller sampling time of water level data (e.g. 10 minute) for reducing the sharp oscillations and giving enough points for producing better local models to handle these oscillations; to implement a mixture of various local models (e.g. ANN) in the phase space which perform the best for predicting future trajectories of a particular condition or regime; to reconstruct the phase space from time series using non-equidistance time delay method which can unfold the attractor better; and to select longer size of cross validation data set.

TABLE 6-3: PERFORMANCE OF UNIVARIATE GLOBAL NEURAL NETWORK MODEL WITH PARAMETER OPTIMIZED BY EXHAUSTIVE SEARCH.

Neural Networks		Prediction horizons				
		1 hrs	3 hrs	6 hrs	10 hrs	12 hrs
Surges						
Stormy	No.hidden	10	7	5	10	9
	RMSE(CV)	10.93	21.30	28.47	35.09	39.43
	RMSE(Test)	**10.72**	19.46	22.34	**30.00**	**31.50**
Non-stormy	No.hidden	6	10	10	7	6
	RMSE(CV)	4.55	6.15	7.09	8.47	8.32
	RMSE(Test)	**5.22**	7.18	**8.09**	**9.46**	**9.45**
Water levels						
Stormy	No.hidden	7	9	1	10	10
	RMSE(CV)	15.07	26.35	39.78	45.81	47.74
	RMSE(Test)	**14.11**	**24.52**	**32.30**	**35.99**	**34.95**
Non-stormy	No.hidden	10	8	8	7	2
	RMSE(CV)	7.51	12.16	14.72	14.76	12.02
	RMSE(Test)	**8.03**	**11.89**	**16.07**	**17.27**	**13.45**

TABLE 6-4: PERFORMANCES OF MULTIVARIATE PREDICTIVE CHAOTIC MODEL AND GLOBAL NEURAL NETWORK FOR STORM SURGE PREDICTION WITH PARAMETERS OPTIMIZED BY EXHAUSTIVE SEARCH.

PHor (hrs)	Surge$_{HvH}$		WL$_{HvH}$		Surge$_{EPF}$		Wind$_{HvH}$		Press$_{HvH}$			Multivariate Predictive chaotic model			Multivariate Neural Networks		
	τ	m	τ	m	τ	m	τ	m	τ	m	k	RMSE (CV)	RMSE (test)	No. hdn	RMSE (CV)	RMSE (test)
Storm Condition																
1	10	8	4	6	1	5	5	5	4	5	13	15.52	8.809	5	6.48	**6.53**
3	10	8	4	6	3	4	2	2	1	4	13	22.21	**11.99**	7	15.87	16.78
6	10	8	4	6	3	5	4	5	5	2	13	37.96	21.12	8	22.24	**20.45**
10	10	8	4	6	3	5	5	5	5	3	13	43.45	31.32	8	28.054	**30.37**
12	10	8	4	6	4	4	3	2	1	2	13	47.28	34.80	8	28.74	**28.09**
Non-storm Condition																
1	10	8	4	6	1	4	2	4	1	5	80	3.40	4.79	5	4.20	**4.76**
3	10	8	4	6	1	2	2	4	2	5	80	2.94	**6.59**	6	7.75	8.69
6	10	8	4	6	1	5	4	2	5	5	80	6.69	**8.05**	2	9.75	11.02
10	10	8	4	6	1	5	5	2	1	2	80	8.02	**10.52**	7	9.96	10.84
12	10	8	4	6	1	4	5	2	5	3	80	6.59	**10.73**	4	10.96	12.37

TABLE 6-5: PERFORMANCES OF MULTIVARIATE PREDICTIVE CHAOTIC MODEL AND GLOBAL NEURAL NETWORK FOR WATER LEVEL PREDICTION WITH PARAMETERS OPTIMIZED BY EXHAUSTIVE SEARCH.

Water Level Prediction																	
PHor (hrs)	$Surge_{HvH}$		WL_{HvH}		WL_{EPF}		$Wind_{HvH}$		$Press_{HvH}$		Multivariate Predictive chaotic model			Multivariate Neural Networks			
	τ	m	τ	m	τ	m	τ	m	τ	m	k	RMSE (CV)	RMSE (test)	No. hdn	RMSE (CV)	RMSE (test)	
Storm Condition																	
1	10	8	4	6	5	4	1	5	3	5	300	13.97	12.44	4	7.94	**8.68**	
3	10	8	4	6	5	2	2	5	2	5	300	44.45	56.99	4	15.42	**15.66**	
6	10	8	4	6	5	2	2	5	3	2	300	100.12	113.25	6	25.93	**27.32**	
10	10	8	4	6	5	3	3	2	1	5	300	79.79	92.25	9	32.17	**29.11**	
12	10	8	4	6	1	4	4	2	3	4	300	61.37	60.05	2	33.89	**31.84**	
Non-storm Condition																	
1	10	8	4	6	2	4	3	2	3	2	300	5.98	7.29	1	3.27	**4.04**	
3	10	8	4	6	5	2	4	5	1	4	300	63.43	57.86	6	6.10	**7.28**	
6	10	8	4	6	4	4	4	5	4	3	300	112.75	113.89	5	10.77	**11.86**	
10	10	8	4	6	5	2	2	2	4	3	300	98.60	87.73	4	12.04	**12.72**	
12	10	8	4	6	5	4	1	2	4	2	300	53.03	48.53	8	9.95	**12.78**	

6.14 K-fold Cross Validation

Cross-validation is a technique for assessing how a predictive model will generalize or perform to an independent data set (Mosteller, 1948). This technique is useful as protection against testing hypotheses suggested by the data (type III errors). The cross-validation involves partitioning a dataset into several complementary sub datasets, performing the analysis on one subset (called the training set), and validating the analysis or prediction on the other subset (called the validation set or testing set). If k-sub datasets are obtained by randomly partitioning the original data set, this process is so-called k-fold cross validation. For reducing the variability, multiple rounds of cross-validation are often performed using different partitions, and the validation results are averaged or combined over the rounds.

In the application for storm surge predictions, it is seen that different size of the validation sub dataset influencing the model evaluation results. Figure 6-21 depicts the chaotic model predictions (3 hours ahead) and observations for storm surges at Hoek van Holland station during stormy period (2160 data points). These 2160 data points can be divided into two groups 1-200 data points (less fluctuated surges) and 750-900 data points (highly fluctuated surges). The model evaluation measures are different depending on the size and the condition of validation data sets used. Table 6-6 shows the 6-folds cross validation errors are obtained for surge data (1990-1995) with one-year validation data set (RMSE). All model errors for 6 sub-datasets of validation are averaged in order to demonstrate the general accuracy of predictive chaotic model for different prediction horizons.

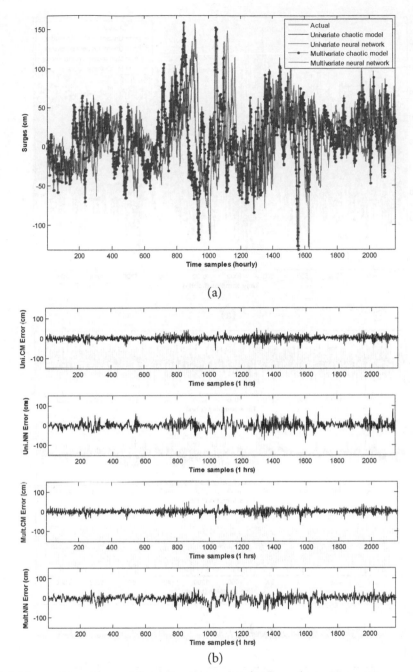

(a)

(b)

Figure 6-19: (a) The comparison of storm surge predictions between univariate and multivariate predictive chaotic models and neural networks at Hoek van Holland during the stormy period (1-Jan-1995 till 31-Mar-1995) based on hourly time series. The prediction horizon is 3 hours. The overall RMSE (b) for univariate CM, univariate Global NN, multivariate CM and multivariate Global NN are 12.91, 19.46, 11.99 and 16.78 cm, respectively.

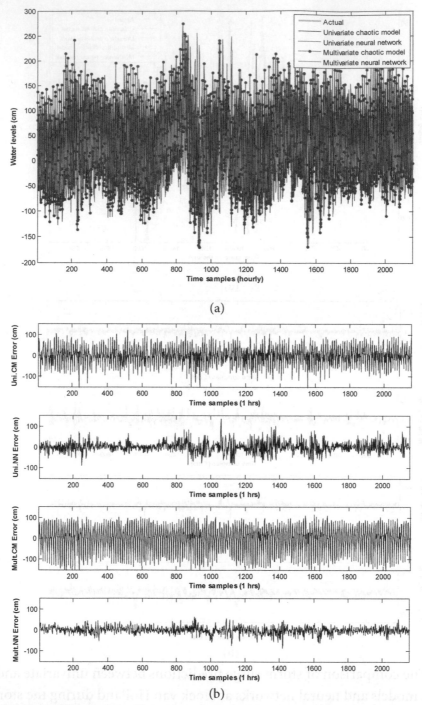

(a)

(b)

Figure 6-20: (a) The comparison of water level predictions between univariate and multivariate predictive chaotic models and neural networks at Hoek van Holland during the stormy period (1-Jan-1995 till 31-Mar-1995) based on hourly time series. The prediction horizon is 3 hours. The overall RMSE (b) for univariate CM, univariate NN, multivariate CM and multivariate NN are 39.20, 24.52, 56.99 and 16.78 cm, respectively.

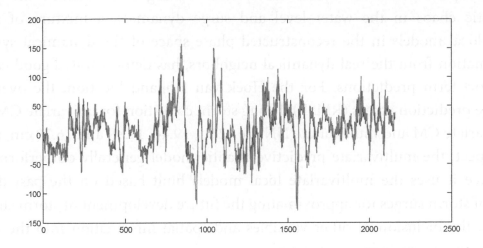

Figure 6-21: Chaotic model predictions (3 hours ahead) for storm surges at Hoek van Holland station during stormy period (2160 data points).

TABLE 6-6: RESULTS OF SEVERAL EVALUATION MEASURES OVER DIFFERENT SIZE OF VALIDATION DATASETS.

Data points	MAE	MSE	RMSE	NRMSE	MAPE
All	8.96	142.57	11.94	0.31	1.60
1-200	7.13	80.42	8.97	0.43	2.16
750-900	11.62	206.24	14.36	0.29	1.16

TABLE 6-7: THE 6-FOLDS CROSS VALIDATION FOR SURGE DATA (1990-1995) WITH ONE-YEAR VALIDATION DATA SET (RMSE).

Validation data set	Prediction horizons				
	1hrs	3hrs	6hrs	10hrs	12hrs
1990	8.32	9.15	16.63	22.69	24.55
1991	7.84	9.02	15.16	20.01	21.41
1992	7.90	8.79	15.51	20.92	22.40
1993	7.65	8.37	15.00	21.33	23.61
1994	7.29	7.72	13.37	18.80	20.93
1995	7.65	8.10	13.92	19.77	22.20
Average	7.78	8.53	14.93	20.59	22.52

6.15 Summary

Based on the nonlinear chaos analysis, the dynamics of both water levels and surges along the Dutch coast can be characterized as deterministic chaos. The presence of the chaotic dynamics together with the positive Lyapunov exponents implies that there are limits of predictability for any model. However, short-term reliable predictions are possible. The chaotic behavior occurs because water levels and surges, including astronomical contributions and the contributions from the meteorological forcing, are the result of a complex, coupled nonlinear dynamical system. Taking into account the presence of deterministic chaos in the water level and surge dynamics, a mixture of multivariate predictive local models in the reconstructed phase-space of the dynamical system, which uses information from the real dynamical neighbors, has demonstrated good capability for reliable short-term predictions. For the Hoek van Holland location, the overall 3 hours ahead surge prediction errors (RMSE) during storm condition for univariate CM, univariate NN, multvariate CM and multivariate NN are 12.91, 19.46, 11.99 and 16.78 cm, respectively. In this respect, the multivariate predictive chaotic model generally outperforms the other models since it uses the multivariate local models built based on the past development similarity of storm surges for approximating the future development of storm surges. This is also due to the inclusion of other variables and spatial information into the multivariate models resulting in the increase of accuracy.

With respect to the model for predicting water levels, it can be seen that it is less accurate than the model for surge since the water level is mainly driven by periodic behavior and this is well predicted by linear system such as Fourier analysis. For improvements, it is recommended to implement a mixture of various local models (like radial basis function), to employ smaller data sampling time and to construct non-equidistance phase space.

It is indicated that the size and condition of data sets chosen as testing or validation are sensitive to the evaluations of model errors. The k-fold cross validation is used for avoiding the type III hypothesis errors by partitioning the surge data (1990-1995) into 6 parts (one-year validation data set). The general errors of predictive chaotic models for different prediction horizons were obtained, for instance 8.53cm errors for 3 hours predictions of one-year validation.

CHAPTER 7: ENHANCEMENTS: RESOLVING ISSUES OF HIGH DIMENSIONALITY, PHASE ERRORS, INCOMPLETENESS AND FALSE NEIGHBORS

"Because of recent improvements in the accuracy of theoretical predictions based on large scale ab initio quantum mechanical calculations, meaningful comparisons between theoretical and experimental findings have become possible."

Yuan T. Lee

This chapter presents several important enhancements for resolving a number of issues associated with high dimensionality of phase space, phase shift prediction errors and presents ways of building a chaotic model from incomplete time series.

7.1   Phase Space Dimensionality Reduction

7.1.1 Introduction

In many cases, not all the measured variables are important for understanding the underlying phenomena. Some measured variables might be irrelevant and the original representation of the data might have redundancies due to the existence of high correlations among the measured variables. Dimensionality reduction techniques can be utilized to remove such redundancies and generate a compact and yet meaningful representation of the original data. Dimensionality reduction can be defined as the process of transforming data residing in a high dimensional space to a low dimensional subspace in such a way that the transformation ensures the maximum possible preservation of information.

This chapter discusses a dimensionality reduction method applied to reduce the dimension of univariate and multivariate time-delayed phase space in order to improve the performance of chaotic model predictions. The reconstructed phase space of a dynamical system normally has a high m-dimensional phase space with τ time delayed coordinates.

This structure may consist of a number of irrelevant and redundant variables even though a suitable pair of embedding dimension m and time delay τ are appropriately selected when performing the phase space reconstruction. The fact of equidistance time-delayed variables in the phase space reconstruction might also induce some redundancies. We propose the phase space dimensionality reduction based on principal component analysis (PCA) to solve these issues by creating a compact and lower dimensional phase space of a dynamical system which can improve the accuracy of chaotic model predictions. Similar researches have been done, such as an attractor reconstruction from univariate time series with a distortion functional comparison of singular system and redundancy criteria studied by Fraser (1989). While Han *et al.* (2006) extracted the feature components of noisy multivariate time series based on singular value decomposition (SVD) and applied ANN for predicting a dynamical system. Yet, to the best of our knowledge, none of papers concentrated on the phase space dimensionality reduction on improving univariate and multivariate chaotic model predictions.

For testing how phase space dimensionality reduction may improve predictive chaotic model performance, both the sea water level and surge time series data along the Dutch coast were considered.

7.1.2 Problems of dimensionality

One of the uses of dimensionality reduction technique is to overcome the curse of dimensionality. The curse of dimensionality describes the problems associated with the rapid (exponential) increase in volume caused by adding extra dimensions to a space. In many cases, not all the measured variables are important for understanding the underlying phenomena. There may be variables whose variance is lesser that the measurement noise, hence these variables are irrelevant. Also, the original representation of the data might have redundancies due to the existence of correlations between the variables. Removing such redundancies can generate a compact and yet meaningful representation of the original data.

Dimensionality reduction can be seen as the process of transforming data residing in a high dimensional space to a low dimensional subspace in such a way that the transformation ensures the maximum possible preservation of information. Dimensionality reduction problems can be formulated as follows: Let $\mathbf{X} = [x_1, x_2, x_3, ..., x_n]$ be a set of n data points in a d-dimensional space, i.e. $x \in \Re^d$, then a dimensionality reduction method tries to find a corresponding output set of patterns $\mathbf{Y} = [y_1, y_2, y_3, .., y_n]$ such that $y_i \in \Re^m$, where $m \ll d$ and $\mathbf{Y}$ provides the most faithful representation of $\mathbf{X}$ in the lower dimensional space.

7.1.3 Principal component analysis

Principal component analysis (PCA) is one of the most popular and widely used techniques for dimensionality reduction that is mathematically defined as an orthogonal linear transformation (Lee & Verleysen, 2007). It transforms the data to a new coordinate system such that the greatest variance by any projection of the data comes to lie on the first coordinate (called the first principal component), the second greatest variance on the second coordinate, and so on. PCA is often presented using the eigenvalue/eigenvector approach of the covariance matrices. For efficient computation, the singular value decomposition (SVD) of the data matrix that is used. Let $\mathbf{Y}$ be a time series data with m records of dimension d. Assume the dataset is mean-centered by making E[$\mathbf{Y}$]=0. An efficient PCA method is calculated based on finding the singular values and orthonormal singular vectors of the $\mathbf{Y}$ matrix, as follows:

$$\mathbf{Y} = \mathbf{U}\Sigma\mathbf{V}^T \tag{7.1}$$

where $\mathbf{U}$ and $\mathbf{V}$ are the left and the right singular vectors of $\mathbf{Y}$, and Σ is a diagonal matrix with positive singular values. Using covariance matrix to calculate the eigenvectors, let $\mathbf{C}$=E[$\mathbf{Y}^T\mathbf{Y}$] represent the covariance matrix of $\mathbf{Y}$. Then the right singular vectors contained in $\mathbf{V}$ of Eq.(5) are the same as those normalized eigenvectors of the covariance matrix $\mathbf{C}$ in Eq.(6). In addition, if the nonzero eigenvalues of $\mathbf{C}$ are arranged in a descending order, then the k-th singular value of $\mathbf{Y}$ is equal to the square root of the k-th eigenvalue of $\mathbf{C}$.

The main procedure of PCA consists of: obtaining some data, subtracting the mean, calculating the covariance matrix, computing the eigenvectors and eigen values of the covariance matrix, choosing components and forming a feature vector and deriving the new data set.

7.1.4 Reducing the phase space dimension

The dimension of the data is the number of variables that are measured on each observation. In predictive chaotic modeling, the dimension refers to phase space dimension constructed from univariate or multivariate time-delayed variables. One of the problems with high-dimensional datasets or space is that, in many cases, not all the measured variables are "important" for understanding the underlying phenomena of interest. Often reducing the dimension of the original data leads to the increased accuracy of the resulting model.

Phase space dimensionality reduction using PCA was implemented in such a way only those characteristics of the time-delayed coordinate variables that contribute most to its variance (the lower-order principal components) were retained. Such low-order components often contain the "most important" aspects of the dynamical system. The phase space dimensional reduction is not only effectively reducing the complexity of dynamical system, but also removing data noises. Furthermore, machine learning methods can also provide a mapping from the high dimensional space to the embedded (reduced) space.

We propose the time-delayed phase space dimensionality reduction using principal component analysis (PCA) in the process of phase space reconstruction. The phase space of a dynamical system is normally reconstructed with a high dimensional space. Let X with size of (m, τ) be a trajectory matrix produced by method of delays with appropriate choice of embedding dimension m and time delay τ. Then, PCA is applied to reduce the phase space dimension of m into lower dimension n. This produces a compact and yet meaningful data representation of the dynamical system. Based on the reduced univariate or multivariate phase space reconstruction, the chaotic model predictions can be developed using the adaptive predictive local models produced based on the dynamical neighbors.

7.1.5 Model results and discussion

Figure 7-1 shows an Eigen values against principle components plot presenting information loss due to 19-dimensional multivariate phase space reduction of Hoek van Holland (HvH) surges time series data consisting of four variables: surges, wind, air pressure at HvH and surges at Euro platform (EPF).

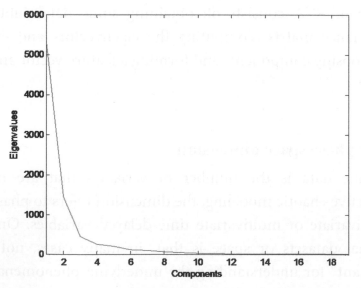

Figure 7-1: Eigen values against principal components plot indicating information loss due to reduction for HvH surges time series data. For this case, 19-dimensional multivariate phase space Y=(HvH(6,6),Wind(1,4),DP(6,3),EPF(1,6)) is reduced to 6 dimensions.

Table 7-1 summarizes the performances of univariate and multivariate predictive chaotic models with and without dimensionality reduction and artificial neural network (ANN) model for non-stormy and stormy conditions. These results show slightly better performance of univariate predictive chaotic models with dimensional reduction for stormy condition compared to the other models including ANN (up to 12 hours ahead).

TABLE 7-1: THE PERFORMANCE COMPARISON OF UNIVARIATE AND MULTIVARIATE ANN AND PREDICTIVE CHAOTIC MODELS WITH AND WITHOUT PHASE SPACE DIMENSIONAL REDUCTION FOR THE SURGE PREDICTION (M=VARIABLE, τ=VARIABLE, K=50-100 FOR NON-STORMY PERIOD AND K=9-100 FOR STORMY PERIOD).

| | RMS Error (cm) for different prediction horizons (1 sample=1 hour) | | | | |
	1 hrs	6 hrs	8 hrs	10 hrs	12 hrs
Non-storm condition					
Univariate ANN	2.10	4.45	5.27	5.54	**5.05**
Univariate predictive chaotic models	2.28	4.73	5.64	6.11	5.18
Univariate predictive chaotic models with PCA	2.28	4.76	5.62	6.09	5.15
Multivariate ANN	6.00	8.61	9.79	11.99	10.38
Multivariate predictive chaotic models	**0.92**	**3.55**	**4.34**	**5.27**	5.44
Multivariate predictive chaotic models with PCA	3.82	5.63	6.33	6.42	6.30
Storm condition					
Univariate ANN	5.07	13.09	15.39	17.88	18.37
Univariate predictive chaotic models	2.62	9.21	11.80	14.06	14.62
Univariate predictive chaotic models with PCA	2.61	**9.18**	**11.76**	**14.01**	**14.61**
Multivariate ANN	9.04	20.05	27.89	30.37	30.57
Multivariate predictive chaotic models	**2.03**	12.16	16.55	20.21	22.82
Multivariate predictive chaotic models with PCA	7.07	13.69	17.00	19.72	22.01

Results for multivariate predictive chaotic models (incorporating information on the meteorological forcing and using additional rules to select dynamical neighbors in the phase space), are mixed. For prediction horizon from 1 to 8 hours PCA does not bring improvement, but for 10 and 12 hours there is some improvement, however also not large.

The main advantage of using PCA is the complexity reduction of the dynamical system especially during storm conditions, and reduction of noise, with no deterioration (and even some gain) in accuracy. It should be mentioned that the experiments with dimensionality reduction should be considered as initial, and we would recommend continuing them, for example in direction of identifying the optimal number of PCA components leading to maximum performance gains.

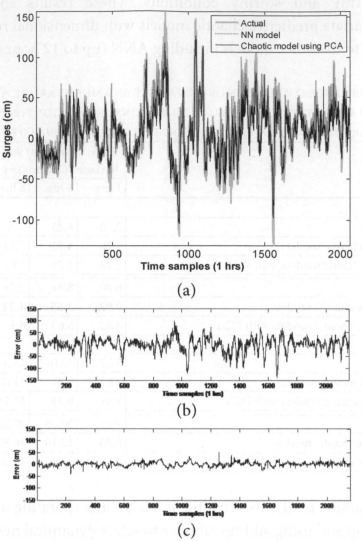

(a)

(b)

(c)

Figure 7-2: (a) Storm surge prediction comparison of multivariate ANN model and univariate predictive chaotic model with phase space dimensionality reduction at Hoek van Holland for the stormy period (1-Jan-1995 till 31-Mar-1995) based on hourly time series. The prediction horizon is 6 hours. The overall RMSE for (b) multivariate ANN is 20.05 cm and (b) for univariate predictive chaotic model with PCA is 9.18 cm.

7.2   Phase Error Correction

7.2.1 Introduction

In chaotic dynamical system, accurate predictions require accurate initial conditions. Although we may have a perfect model and an infinitely long time series of observations, it may be impossible to precisely determine the initial state of a system. The small errors in the initial conditions can be amplified exponentially and this limits on the ability to accurately predict the future states. One of dominant errors is a systematic phase error between model prediction and actual observations. Advanced data assimilation techniques can reduce the predictive chaotic model error by combining the prediction with observations ((Kalnay, 2003; Dercole & Rinaldi, 2008)).

The phase prediction error of a predictive chaotic model can be characterized as phase synchronization (Pikovsky *et al.*, 2002). Phase synchronization is usually applied to two waveforms of the same frequency with identical phase angles with each cycle. However it can be applied if there is an integer relationship of frequency, such that the cyclic signals share a repeating sequence of phase angles over consecutive cycles. The observations as the natural phenomena and the predictive chaotic model built from observables can be described as two chaotic systems or oscillators. They oscillate with a repeating sequence of relative phase angles. Phase synchronization can be performed to eliminate the phase error.

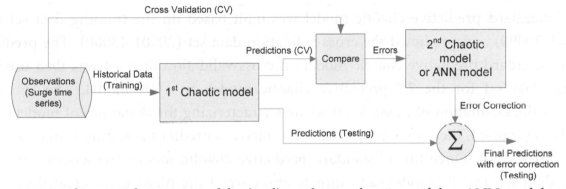

Figure 7-3: A schematic description of the (2nd) predictive chaotic model or ANN model used for correcting the phase error of the (1st) standard chaotic model predictions.

In this work, however, we do not improve the initial conditions, model parameters or model physics, but utilize a specialized model to estimate and correct the prediction errors of predictive chaotic model directly. Two models, ANN model and predictive chaotic model are utilized and trained on the historical phase error data generated by the standard chaotic model predictions. Figure 7-3 schematizes the automated correction of phase error resulted from the chaotic model predictions. The phase error is estimated and corrected by the 2nd model predictive chaotic model or ANN model which is trained based on the 1st standard chaotic model prediction error on cross validation data set. The final prediction on (out-of-sample) testing dataset is performed by these two models by summing up the 1st chaotic model predictions and the error correction estimated by the 2nd model.

7.2.2 Data description

The data set used in previous experiments is used here as well: sea water level, surge, atmospheric pressure and wind speed/direction time series data from seven coastal stations along the Dutch coast; each time series begins January 1st, 1990 and is available until March 31st, 1996, which results in 337249 continuous samples in total for the 10 min time (54768 for the hourly averaged time series). Predictions are made for the Hoek van Holland (HvH). The surge data are split into training, cross-validation (CV) and testing data sets for the 1st and 2nd models (see Table 7-2).

TABLE 7-2: DATA SEPARATION OF HOURLY SURGE DATA FOR TRAINING, VALIDATION AND TESTING DATA SETS.

Time Index	First Model (Standard Predictive chaotic model)			Second Model (Predictive chaotic model or ANN model)	
	Train	CV	Test	Train	Test
Start	1	20001	43001	20001	43001
End	20000	43000	45160	43000	45160

7.2.3 Setting up the 1st standard predictive chaotic model

The 1st standard predictive chaotic model was built based on the training data set (time index=1-20000) and predicted the cross-validation data set (20001-43000). The prediction error of standard predictive chaotic model on cross-validation data set was then used as a training data set for the 2nd predictive chaotic model or ANN model. The 2nd model performs the estimation of possible errors by characterizing the dynamics of model errors. After being trained, these two models can be utilized to predict the testing (unknown) data set (43001-45160). Once the 1st standard predictive chaotic model predicts the surges on testing data set, the 2nd model can estimate and correct the phase error created by the 1st predictive chaotic model. The correction is made by simply summing up these two model predictions. The prediction horizons are 1, 3, 6, 10 and 12 hours for surges at HvH. The following sections describe the nonlinear analysis of surge time series to find the proper values of τ and m (please refer to (Sivakumar, 2004) for more detailed explanation).

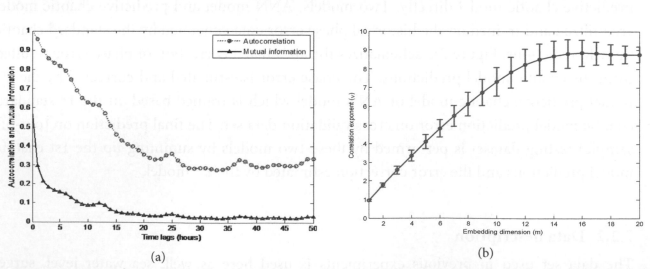

Figure 7-4: (a) Autocorrelation function and mutual information as a function of time lags; (b) Relationship between the correlation exponent τ and embedding dimension m.

7.2.3.1 Finding the proper time delay

The straightforward choice of τ is usually made with the help of the zero-crossing autocorrelation function. Figure 7-4 (a) shows the autocorrelation and mutual information of the surges at HvH. The first minimum value of mutual information is 10 hours for surges.

7.2.3.2 Estimating the appropriate embedding dimension

A proper embedding dimension has to be identified, such that the structure of the attractor becomes invariant. The most widely used fractal dimension quantifier is the correlation dimension d_c (see (Grassberger & Procaccia, 1983b)). Figure 7-5 (b) illustrates that the correlation exponent increases with an increase of the embedded dimension up to a certain value and further saturates. The saturation value of the correlation exponents/dimensions using the optimal time delay of 10 hours is 8.5.

In conclusion, the nonlinear analysis of surge time series recommends the appropriate values of τ and m are 10 and 8, respectively, so the optimal phase space structure for predicting surges during storm condition is:

$$\mathbf{Y}_{t+T} = \left\{ s_t^{hvh}, s_{t-10}^{hvh}, \ldots, s_{t-70}^{hvh} \right\} \qquad (7.2)$$

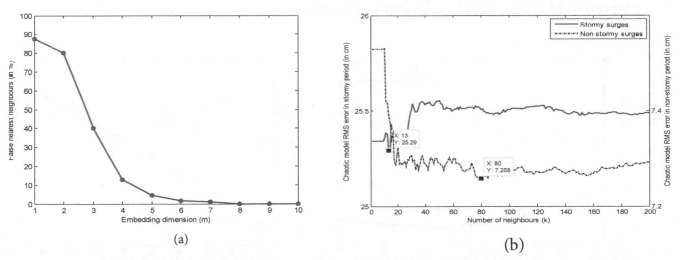

(a)   (b)

Figure 7-5: (a) Percentage of false nearest neighbors; (b) Six-hours ahead prediction error of the predictive chaotic models as a function of the number of neighbors (k) with τ=10 and m=8.

7.2.3.3 Using the proper number of neighbors

Performance based optimization is utilized to find an appropriate number of neighbors (k) by setting up the predictive chaotic model parameters for the surges with τ=10 and m=8 and the number of neighbors (k) run from 1 to 2000. The 3rd-order polynomial local model is used as a reference. Figure 7-5(b) depicts six-hours prediction RMS errors of the predictive chaotic models as a function of the number of neighbors (k). It is seen that the suitable number of neighbors for predicting surges during storm condition is 13.

7.2.4 Setting up the 2nd model (predictive chaotic model and ANN model

7.2.4.1 Predictive chaotic model

Figure 7-6 (a) shows the autocorrelation function and mutual information of the predictive chaotic model errors. It is seen that the mutual information decreases rapidly at time lag 1 hour and does not much change for further time lags. Thus, the optimal value of τ is set to be 1 hour. Figure 7-6(b) draws the correlation exponents as a function of m. The saturation point (correlation dimension) cannot be determined by this plot. Consequently, we executed the exhaustive search to find better estimation of m and obtained the optimal value of $m=12$. This means that the 2nd predictive chaotic model can be built using the phase space structure:

$$\mathbf{Y}_{t+T} = \{s_t^{hvh}, s_{t-1}^{hvh}, ..., s_{t-11}^{hvh}\} \tag{7.3}$$

The prediction horizon of this model is 1 hour ahead. The (one hour ahead) future state is determined by local model based on the dynamical neighbors whose behavior is similar to the last 12 observations. This result could be associated with the most dominant tidal component M2 whose tidal cycle about 12 hours.

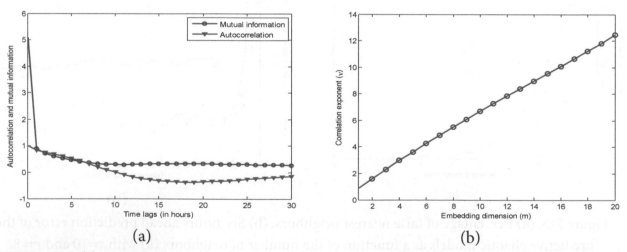

Figure 7-6: (a) Autocorrelation and mutual information and (b) the relationship between the correlation exponent ν and embedding dimension m of the predictive chaotic model errors.

7.2.4.2 ANN model

Backpropagation multi-layer perceptron (MLP) with Levenberg-Marquardt training rule (Haykin, 1999) was utilized and trained using the same input structure as the 2nd predictive chaotic model inputs (Eq.2). The number of hidden neurons of ANN was selected using the exhaustive search in the range [1~10] and we found that three is the optimal number of hidden nodes.

7.2.5 Model results and discussion

The model prediction performance (RMS errors) is summarized in Table 7-3. The use of the 2nd model is able to reduce significantly the presence of phase errors. The 2nd predictive chaotic model outperforms for short-term predictions, whereas the ANN model outperforms for long-term predictions. This is due to the fact that the predictive chaotic model is very sensitivity to initial conditions. In contrast, ANN model is less sensitive to initial condition allowing for more accurate long-term predictions.

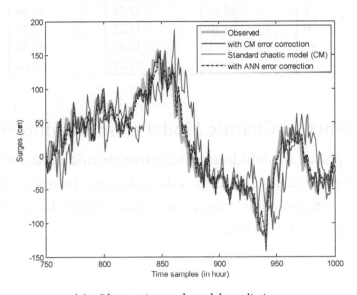

(a) Observation and model predictions

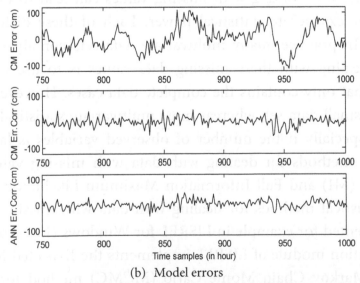

(b) Model errors

Figure 7-7: The observed and predicted surges by standard predictive chaotic model, with error correction by predictive chaotic model and ANN model, and the model prediction errors.

Figure 7-7 shows the observed and predicted surges of the 1st standard predictive chaotic model, with error correction by predictive chaotic model and ANN model during stormy condition at HvH and their model prediction errors. It is seen that the phase errors can be well corrected by predictive chaotic model or ANN model. The approach of using two models enhances the predictability of predictive chaotic model for longer-time prediction horizon.

Table 7-3: Performances (RMS errors) of standard predictive chaotic model with and without error correction using ANN model for stormy condition.

Pred. horiz.	Standard predictive chaotic model	With error correction	
		Predictive chaotic model	ANN model
1	21.33	11.42	18.33
3	24.69	12.74	16.06
6	29.10	13.77	14.75
10	37.02	14.43	12.56
12	39.46	14.81	12.90

7.3   Building Predictive Chaotic Model from Incomplete Time Series

Several methods are explored for building a predictive chaotic model from incomplete time series comprising of: weighted sum of linear interpolations, Bayesian principle component analysis, cubic spline interpolation. Since the data might have different time scale, interpolating and averaging are required.

7.3.1 Introduction

Traditional approaches for working with missing values can lead to biased estimates and may either reduce or exaggerate statistical power. Each of these distortions can lead to invalid conclusions. In practice, many multivariate data sets contain missing values. The traditional way of dealing with these missing data values is to use list wise deletion to generate a data set that only contains the complete data cases. However, list wise deletion may result in a very small data set, whereas most multivariate statistical methods require a large sample size, especially if the number of observed variables is large. Consequently, alternative statistical methods for dealing with data with missing values are of interest. Multiple Imputation (MI) and Full Information Maximum Likelihood (FIML) estimation are two popular statistical methods for dealing with data with missing values. Both these methods are implemented for example in LISREL for Windows (Jöreskog & Sörbom 2005). The Multiple Imputation module of LISREL implements the Expected Maximization (EM) algorithm hnd the Markov Chain Monte Carlo (MCMC) method for imputing missing values in multivariate data sets. Technical details of these methods are available in Schafer

(1997) and Du Toit & Du Toit (2001). Supplementary notes on these methods are also provided by Du Toit & Mels (2002).

Some techniques on how to build predictive chaotic model from incomplete time series were introduced due to the facts that the observed data (including prediction data) are not the same in some aspects, like: time step, prediction horizon, prediction data resolution, time availability of data. Some operational storm surge models do not provide predictions for some locations or some units (i.e. water level but no surge). Missing data is also major issue due to measurement or transmission failures while building predictive chaotic model requires a sequential complete data. Several imputing techniques to solve these issues are described in the following sections.

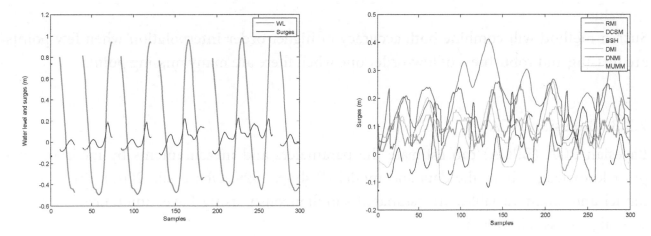

Figure 7-8: Missing observed data and predictions from operational storm surge model with some missing values for Hoek van Holland station

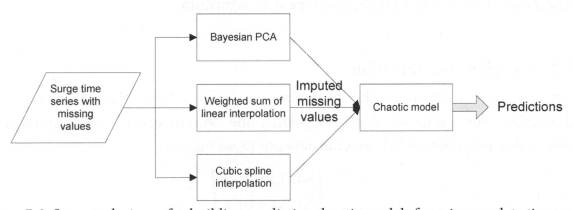

Figure 7-9: Some techniques for building predictive chaotic models from incomplete time series

7.3.2 Weighted sum of linear interpolations

Weighted sum of linear interpolations make several low-order estimates of the missing values from available nearest points from both rows and columns (Pankratov, 1995). The procedure is as follows:

a) Use 10 estimates - 5 from rows and 5 from columns (3 two-point (linear) interpolations left-left, left-right, right-right) and 2 one-point nearest neighbor values

b) Combine these estimates with different weights

c) These weights will reflect availability of these basis points and distance from them

d) They can be tuned so that for the best case (isolated missing points away from the edges) the result will be equivalent to average of 4-point (cubic) interpolations from rows and columns

e) For the worst case (only 1 point is available) the whole matrix will be filled with this value

Such a method will combine both accuracy of higher order interpolation when few points are missing and robustness of low-order one when there are many missing points.

7.3.3 Bayesian PCA

Bayesian PCA (BPCA) can estimate the parameters and measurements by incorporating prior knowledge about the data and model (Bishop, 1999; Oba et al., 2003). Probabilistic model and latent variables are estimated simultaneously using Bayes inference. There are several processes in BPCA:

a) Principle component (PC) regression

b) Bayesian estimation

c) Expectation-maximization (EM)-like repetitive algorithm

7.3.4 Cubic spline interpolation

A form of interpolation where the interpolant is a special type of piecewise polynomial called a spline. For a data set x_i of $n+1$ points, one can construct a cubic spline with n piecewise cubic polynomials between the data points, as follows:

$$S(x) = \begin{cases} S_0(x), x \in [x_0, x_1] \\ S_1(x), x \in [x_1, x_2] \\ \dots \\ S_n(x), x \in [x_{n-1}, x_n] \end{cases} \tag{7.4}$$

where S represents the spline function interpolating the function f, one require:

a) the interpolating property, $S(x_i)=f(x_i)$

b) the splines to join up, $S_{i-1}(x_i) = S_i(x_i)$, i=1,...,n-1

c) twice continuous differentiable, $S'_{i-1}(x_i) = S'_i(x_i)$ and $S''_{i-1}(x_i) = S''_i(x_i)$, $i=1,...,n-1$.

7.3.5 Model results and discussion

Building chaotic model from observed time series with missing values has been introduced by some imputing techniques (weighed sum of linear interpolation, Bayesian PCA, cubic spline interpolation) in phase space. We tested these techniques for predicting storm surges at HVH with 1, 5, 10 and 30% missing values. Chaotic model built from incomplete time series with several imputation techniques has comparable prediction accuracy to the one built from complete time series (see Table 7-4, Table 7-5 and Table 7-6). This demonstrates that the imputing techniques used can be incorporated into the chaotic model for handling the missing values in real operation.

Figure 7-10, Figure 7-11 and Figure 7-12 depict the estimations of missing values by imputing techniques: weighted sum of linear interpolation, Bayesian PCA and cubic spline interpolation, respectively. The blue line denotes the positions of missing values in the time series and these are estimated by imputing techniques. The visualization inspection shows that the cubic spline interpolation provides better estimations of missing values than weighted sum of linear interpolation and Bayesian PCA. Their performances are clearly distinguishable for water level time series, but they have comparable performance for surge time series.

TABLE 7-4: PERFORMANCES (RMS ERRORS) OF CHAOTIC MODEL WITH VARIOUS PERCENTAGES OF MISSING VALUES IMPUTED BY WEIGHTED SUM OF LINEAR INTERPOLATION.

PHoriz.	Missing values				
	0%	1%	5%	10%	30%
1	10.7	**10.7**	**10.8**	11.8	15.3
3	11.0	11.1	**11.3**	12.2	15.3
6	21.0	**21.0**	**21.0**	20.9	**21.0**

TABLE 7-5: PERFORMANCES (RMS ERRORS) OF CHAOTIC MODEL WITH VARIOUS PERCENTAGES OF MISSING VALUES IMPUTED BY BAYESIAN PCA.

PHoriz.	Missing values				
	0%	1%	5%	10%	30%
1	10.7	10.8	12.2	**11.1**	12.2
3	11.0	11.1	12.9	**11.2**	12.9
6	21.0	**21.0**	21.1	21.1	**21.0**

TABLE 7-6: PERFORMANCES (RMS ERRORS) OF CHAOTIC MODEL WITH VARIOUS PERCENTAGES OF MISSING VALUES IMPUTED BY CUBIC SPLINE INTERPOLATION.

PHoriz.	Missing values				
	0%	1%	5%	10%	30%
1	10.7	**10.7**	11.5	11.6	**11.8**
3	11.0	**11.0**	11.6	11.7	**12.0**
6	21.0	21.1	21.2	**20.8**	22.0

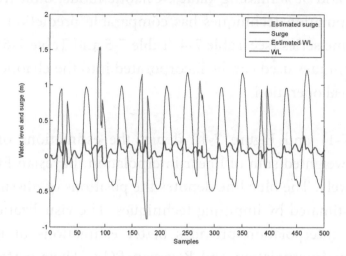

Figure 7-10: Estimation of missing water level and surge in phase space using weighted sum of linear interpolation from nearest neighbors in comparison with the actual observation.

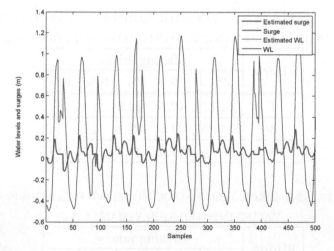

Figure 7-11: Estimation of missing water level and surge in phase space using Bayesian PCA in comparison with the actual observation.

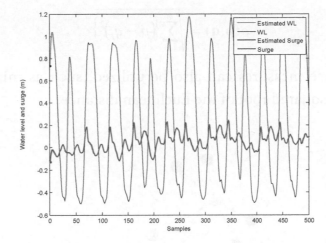

Figure 7-12: Estimation of missing water level and surge using cubic spline interpolation in comparison with the actual observation.

7.4   Finding True Neighbors

The true neighbors here means that these neighbors have the similar dynamical characteristics or properties (i.e. similar type of storm development) to the reference or actual points in data sets. A searching algorithm may find false neighbors that do not have similar dynamical characteristics; however they are selected as neighbors by the algorithm.

In this research, two different methods are used to find true neighbors in the phase space aiming to build local models for predicting storm surges. The first method employs the standard method based on Euclidean distance to find neighbors. The second one utilizes the new searching method by employing the principles of vectors (scalar and its direction) to find neighbors along the approximated trajectory in the reconstructed phase space. Finding true neighbors can be improved by means of this new searching method.

7.4.1 Euclidean distance method

The first method used to find the neighbors in phase space is based on the Euclidean distance between two points in m-dimensional space. The distance is measured from the actual point to all other points in the phase space. The actual point refers to the last data point in training data set before prediction. The neighbors are chosen if the distances of neighboring points are less than the threshold value. The criteria of distance threshold can also be replaced with the number of neighbors. The number of neighbors technique seeks the nearest neighbors as many as the pre-defined number of neighbors (Figure 7-13). Let $\mathbf{p}=(p_1, p_2, ..., p_m)$ and $\mathbf{q}=(q_1, q_2, ..., q_m)$ be two points in m-dimensional space, the Euclidean distance (d) between them can be calculated as:

$$d(\mathbf{p}, \mathbf{q}) = \sqrt{\sum_{i=1}^{m} \left\{ (p_i - q_i)^2 \right\}} \qquad (7.5)$$

Other Euclidean distance measures can also be utilized, such as minimum or maximum Euclidean distance and squared root of the Euclidean distance.

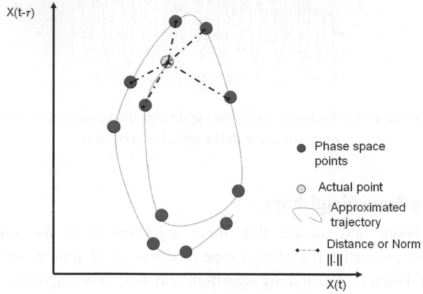

Figure 7-13: The searching of neighbors using Euclidean distance method in 2-dimensional phase space.

It is possible to introduce clustering strategy in the neighbor searching processes. The clustering procedure is performed by grouping the points along the trajectory in m-dimensional phase space according into a pre-defined number of clusters. Technically, this is done by sorting the phase space matrix, taking a reference vector $\mathbf{x}_t$ from the phase space matrix and estimating the centre of each cluster. Using the established clusters, the neighbors of an actual point are searched by finding the nearest cluster (center) to the actual point within a certain distance threshold value and searching the neighboring points inside the selected clusters.

7.4.2 The new trajectory based method

The trajectory based method arises from the main idea that finding true neighbors does not only depend on the distance between two points in the m-dimensional phase space, but also the distance and direction of two different trajectories (sequences of points in phase space) partly formed by these two points. Technically, a point in phase space is an m-dimensional vector. A trajectory for measuring distance here consists of a connection of at least two points in phase space, which has scalar value and its direction. The trajectory matrix can be created from a phase space matrix depending on how many points to create a trajectory.

The trajectory neighbors are obtained by searching for trajectories which are nearest distance and similar direction to the actual trajectory (a trajectory formed by the last point). These trajectory neighbors contain points that can used to build the predictive local models. The principle of trajectory based method for finding true neighbors is illustrated in Figure 7-15.

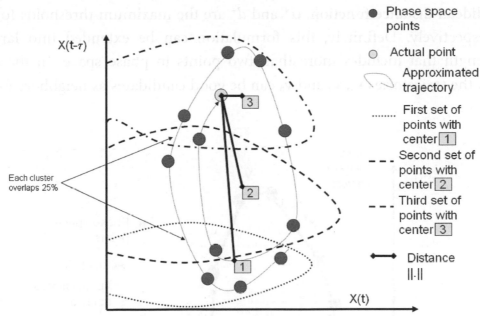

Figure 7-14: The searching of neighbors using Euclidean distance method and clustering strategy in 2-dimensional phase space.

Let $\mathbf{v}_1$ and $\mathbf{v}_2$ be two trajectories lines represented as vectors with particular directions in a m-dimensional phase (Euclidean) space Γ. The angle (α) between these two trajectories is given by inverse cosine of the dot product between two (normalized) vectors ($\mathbf{v}_1$ and $\mathbf{v}_2$), expressed as:

$$\alpha = \arccos\left(\frac{\mathbf{v}_1 \bullet \mathbf{v}_2}{\|\mathbf{v}_1\|\|\mathbf{v}_2\|}\right) \tag{7.6}$$

where the angle α is in the range of $[0°, 180°]$.

Suppose the vector $\mathbf{v}_1$ be a trajectory that contains the actual point and is formed by two points in m-dimensional phase space, the searching for the trajectory neighbors is performed by calculating the angles and distances between trajectory $\mathbf{v}_1$ and the other trajectories given by a set of (n-1) angles $\Psi=(\alpha_{v1v2}, \alpha_{v1v3}, ..., \alpha_{v1vn})$ and (n-1) distances $\Omega=(d_{v1v2}, d_{v1v3}, ..., d_{v1vn})$, where n is the total number of trajectory sections in phase space. The trajectories are selected as good neighbor candidates if they are near distance and similar angle (direction) to the actual trajectory $\mathbf{v}_1$. Mathematically, the trajectory neighbors (TN) can be formulated as:

$$TN = \left\{ \forall (\mathbf{v}_1, \mathbf{v}_2) \in \Gamma^2, \exists (\mathbf{v}_1, \mathbf{v}_2), \alpha^* \in \Psi, d^* \in \Omega \middle| 0 < \arccos\left(\frac{\mathbf{v}_1 \bullet \mathbf{v}_2}{\|\mathbf{v}_1\|\|\mathbf{v}_2\|}\right) \leq \alpha^* \wedge 0 < d(\mathbf{v}_1, \mathbf{v}_2) \leq d^* \right\} \quad (7.7)$$

where $\mathbf{v}_1$ and $\mathbf{v}_2$ are any two trajectories formed by two points in m-dimensional phase space Γ^2, d is Euclidean distance function, α^* and d^* are the maximum thresholds for angle and distance, respectively. Definitely, this formulation can be extended into larger size of trajectory length that includes more than two points in phase space. In the example of Figure 7-15, the trajectories $\mathbf{v}_4$, $\mathbf{v}_5$ and $\mathbf{v}_6$ can be good candidates as neighbors for trajectory $\mathbf{v}_1$.

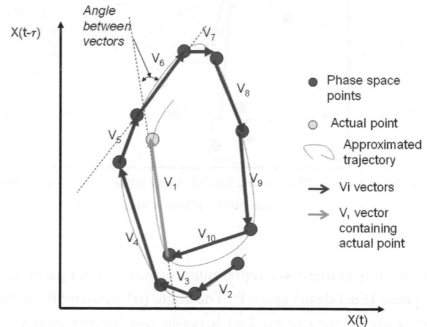

Figure 7-15: The searching of neighbors using trajectory based method in 2-dimensional phase space. Trajectories $\mathbf{v}_4$, $\mathbf{v}_5$ and $\mathbf{v}_6$ can be good candidates as true neighbors for trajectory $\mathbf{v}_1$.

7.4.3 Model results and discussion

The values of time delay, embedding dimension and maximum number of neighbors were set to 19 hours, 20 dimension and 300 neighbors, respectively. Six experiments (Y1 till Y6) were defined with different configurations of parameter settings for building predictive chaotic model (1 hour ahead). The first experiment (Y1) utilizes the clustering strategy with 3000 maximum points in each cluster, cubic function as local model, Euclidean distance method with descending sorting, and finally the number of maximum neighbors to construct the local model is 300. The second configuration (Y2) of parameters differs from the previous one in the way the program order the selected neighbors, in this case is ascending Euclidean distance method. The third one is similar than Y2 except clustering is not used. The fourth and fifth experiments (Y4 and Y5) differ from the Y3 in the way the neighbors are found using trajectory based method with a selection criterion that the angle was set in a range between the maximum value and 70% of the maximum value (not an ideal parameter setting). The Y5 experiment used 300 neighbors whereas Y4 experiment utilized variable number of neighbors. The last experiment (Y6) differs from the Y5 in the criteria of neighbor selection at which the angle should be in between zero and 90°.

TABLE 7-7 PERFORMANCE COMPARISON OF PREDICTIVE CHAOTIC MODELS WITH DIFFERENT PARAMETER SETTINGS (TRAJECTORY BASED METHOD AND CLUSTERING STRATEGY).

Parameters	Y1	Y2	Y3	Y4	Y5	Y6
τ	19	19	19	19	19	19
m	20	20	20	20	20	20
k_{min}	50	50	50	50	variable	variable
k_{max}	300	300	300	300	variable	variable
Clustering	cluster	cluster	no cluster	no cluster	no cluster	no cluster
Max points in a cluster	3000	3000	3000	3000	3000	3000
Local model	cubic	cubic	cubic	Cubic	cubic	cubic
Searching method	Euclidean	Euclidean	Euclidean	trajectory	trajectory	trajectory
Prediction horizon	6 hours	6 hours	6 hours	6 hours	6 hours	6 hours
Theiler's window	19 steps	19 steps	19 steps	19 steps	19 steps	19 steps
Sorting method	descending	ascending	ascending	descending	descending	descending
MAE	2.1	2.7	2.8	3.6	4.1	1.1
MSE	6.6	12.9	13.5	20.8	30.8	1.8
RMSE	2.6	3.6	3.7	4.6	5.6	1.3
NRMSE	0.1	0.1	0.1	0.1	0.1	0.0
MAPE	0.5	0.4	0.3	0.8	0.5	0.5

The model errors of Y1 till Y6 are listed in Table 7-7. There is a significant improvement in the accuracy of predictive chaotic model if the trajectory based method is used. In case of Y4 experiment, the high accuracy of predictive chaotic model is not expected if the trajectory based method with angle restriction (less than the maximum angle and greater than 70% of this value) is used. However, a considerable improvement was obtained when trajectory based method was utilized with angle setting in a range between 0 and 90°. The experiment Y6 with the trajectory based method has improved prediction accuracy with 1.3cm RMS error whereas the Y1 with Euclidean based method and clustering strategy has 2.6cm RMS error. A plot comparing the predictive performance of the Y1 and Y6 models is depicted in Figure 7-16. This plot illustrates that the model predictions agree with the observations for descending Euclidean distance method with clustering and trajectory based method with angle between 0 and 90°.

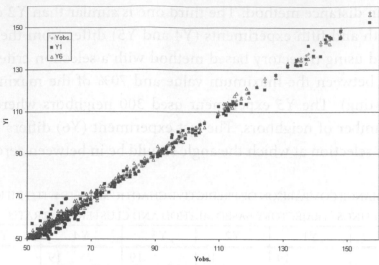

Figure 7-16: Relationships between observed (Y_{obs}) and predicted values of the Y1 model (Euclidean based method with clustering) and Y6 model (trajectory based method).

7.5 Summary

Several enhancements are considered. First, the method of phase space reconstruction of a dynamical system incorporating dimensionality reduction is presented. Based on this reduced phase space, multivariate predictive local models are built from the real dynamical neighbors. Both water levels and surges time series along the Dutch coast which can be characterized as deterministic chaos were taken for testing our models. The results have shown that the use of dimensionality reduction method in the phase space reconstruction can improve the performance of univariate and multivariate predictive chaotic models outperforming ANN. For the Hoek van Holland location, the overall prediction error for surges 10 hours ahead is about 5 cm and 14.5 cm for non-storm and storm conditions respectively, which are well comparable with the physically-based numerical model WAQUA/DCSM currently being used in practice. This demonstrates that the proposed model can serve as an efficient tool for accurate and reliable short-term predictions of water levels and surges to support decision-makers in ship navigation and flood prediction.

Second, an approach of correcting phase error in the chaotic model predictions is presented. In the considered application to storm surge modeling, the predictive chaotic model and ANN model can identify the dynamical behavior of the phase error of a standard predictive chaotic model and is able to estimate and correct it. This demonstrates that the proposed technique can enhance the predictability of a predictive chaotic model for longer-term predictions.

Third, chaotic model built from incomplete time series with imputation has comparable prediction accuracy to the one built from complete time series. This demonstrates that the imputing techniques used can be incorporated into the chaotic model for handling the missing values in the real operation.

Fourth, a novel approach for solving the issue on false neighbors, so-called trajectory based method is introduced. The main idea of this method arises from the fact that finding true neighbors does not only depend on the distance between two points in the m-dimensional phase space, but also the distance and direction of two different trajectories partly formed by these two points. This technique allows for improving the accuracy of predictive chaotic model – due the fact that the predictive local models are constructed using more true neighbors and less false neighbors.

CHAPTER 8: COMPUTATIONAL INTELLIGENCE IN IDENTIFYING OPTIMAL PREDICTIVE CHAOTIC MODEL

"As natural selection works solely by and for the good of each being, all corporeal and mental endowments will tend to progress toward perfection."

Charles Darwin

This chapter describes the ways of optimizing predictive chaotic model parameters leading to more accurate predictions. Three methods, grid search, genetic algorithms and adaptive cluster covering, are discussed and implemented for optimizing predictive chaotic model.

8.1 Introduction

The characteristics of the strange attractors of a chaotic system can be analyzed by sampling a part of the output chaotic time series of a system. The method that is commonly used is the state space reconstruction in delay coordinate proposed by Packard *et al.* (1980). Further, Floris Takens introduced his famous Taken's theorem stating that the unstable periodic obits (strange attractor) could be recovered properly in an embedding space whenever a suitable embedding dimension $m \geq 2d+1$ (d is the dimension of chaotic system) is detected; that is, the orbits in the reconstructed space R^m keeps a differential homeomorphism with the original system (Takens, 1981).

It is very important to select a suitable pair of embedding dimension m and time delay τ when performing the phase space reconstruction. The precision of τ and m is directly related with the accuracy of the invariables of the described characteristics of the strange attractors in phase space reconstruction. There exist two different points of view for doing this.

The first one is that m and τ are not correlated with each other; that is, m and τ can be selected independently (Takens has proved that m and τ are independent in a chaotic time series with infinite length and no noise). Under this rule, a commonly used approach, GrassbergerProcaccia algorithm, for calculating the embedding dimension m was proposed by Albano *et al.* (1988). For the time delay τ, there are three criteria for its selection: (1) time series correlation approaches: autocorrelation, mutual information (Fraser, 1989), high-order correlations (Albano *et al.*, 1991) (2) approaches of phase space extension: wavering technique (Buzug & Pfister, 1992) and average displacement (Rosenstein *et al.*, 1994), and (3) multiple autocorrelation and non-bias multiple autocorrelation (Jiayu *et al.*, 2006).

The second viewpoint is that m and τ are closely related because the time series in the real world could not be infinitely long and could hardly avoid being noised. A great deal of experiments indicate that m and τ tie tightly up with the time window $t_w=(m-1)\tau$ for the reconstruction of the phase space. For a given chaotic time series, t_w is relatively steadfast. An irrelevant partnership between m and τ will directly impact the equivalence between the original system and the reconstructed phase space. Therefore, the combination approaches for computing m and τ accordingly come into being, e.g., small-window solution (Kugiumtzis, 1996), C–C method (Kim et al., 1999), and automated embedding (Otani and Jones, 2000). Most researchers consider that the second viewpoint is more practical and reasonable in the engineering practice than the first one.

The research on combination algorithm for optimal embedding dimension and delay time is of interest in improving the predictive chaotic model performance. Moreover, an automated embedding algorithm which estimates a near optimum embedding dimension and delay time can be developed.

In addition to optimizing the time delay (τ) and embedding dimension (m), the non-fixed number of neighbors k [*min,max*] of a dynamical chaos is essential for chaotic model prediction. The optimization of τ and m can be done using an exhaustive search method, but with addition of optimizing parameter k, which is a flexible number of neighbors within the range [*min,max*], then the optimization problem becomes more complex and needs more powerful techniques. Several optimization methods like particle swarm optimization, hierarchical GA, NSGA-II, ant colony optimization can be used to solve this problem (Figure 8-1). This research is extended to the optimization of the phase space structure using dimensional reduction methods with a certain optimality measure. This task also applies to multivariate phase space reconstruction (Murcia, 2009).

The problem of optimization of a predictive chaotic model can be formulated as:

Let, the optimization problem turns out to be:

Minimize   :   error [$M_c(\tau, m, k)$] over training/cross-validation data set

Subject to   : $m_{min} \leq m \leq m_{max},$

$\tau_{min} \leq \tau \leq \tau_{max},$ (8.1)

$k_{min} \leq k \leq k_{max}$

$m, \tau, k \in \mathbb{N}_1$

where M_c is a predictive chaotic model; τ, m, k are time delay, embedding dimension and number of neighbors, respectively; and τ_{min} τ_{min} m_{min} m_{min} are the searching space for time delay and embedding dimension. The objective function is to minimize the predictive chaotic model errors over cross-validation data set. A number of error measures can be used, however in this work, the root mean squared error (RMSE) is applied.

8.2   Randomized Search Algorithms

8.2.1 Grid search

In discrete problems in which no efficient solution method is known, it might be necessary to test a set of possibilities sequentially in order to find the optimal solution, for instance the optimal model parameters. In case of predictive chaotic model, three important parameters (τ, m, k) as decision variables are required to be estimated properly. One of optimization techniques used for this purpose is simply search on a grid, and can be called Grid Search (GS). This technique performs exhaustive search through a subset of parameter space or combinations of decision variables in a certain range by evaluating the objective function and choosing the optimal solution corresponding to the minimum or maximum of the objective function.

8.2.2 Genetic algorithm (GA)

Genetic algorithm (GA) is a method for solving both constrained and unconstrained optimization problems that is based on natural selection, the process that drives biological evolution (Holland, 1975). The genetic algorithm repeatedly modifies a population of individual solutions (e.g. τ, m and k). In the each iteration, the genetic algorithm selects individuals at random from the current population to be parents and uses them to produce new children for the next generation (see Figure 8-1). Over successive generations, the population evolves toward a better solution. It can be applied to solve problems in which the objective function is not continuous, stochastic, or highly nonlinear.

The genetic algorithm uses three main types of processes with specific rules at each step to create the next generation from the current population. Selection rules select the individuals, called parents that contribute to the population at the next generation. Crossover rules combine two parents according with certain rule to form better children for the next generation. Mutation rules apply random changes to individual parents to form new children.

Application of GA to optimize time delay and embedding dimension for a sinusoidal time series was carried out by Vitrano & Povinelli (2001). A binary-string genetic algorithm (GA) was used to reach for the dimensionality and individual delay values for an embedding that better fits a given criterion – in this case, the minimum standard deviation of estimates of the radius of the attractor compared to the mean of those values. The results showed that the GA appears to be a viable method for identifying appropriate time-delay and embedding dimension for certain types of attractors. The technique worked well if the attractor has a relatively uniform radius in phase space.

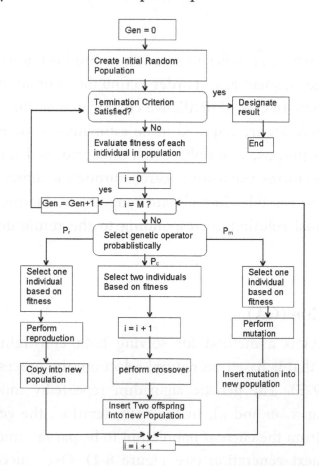

Figure 8-1: Genetic algorithm used for optimizing predictive chaotic model parameters: time delay, embedding dimension and number of neighbors.

8.2.3 Adaptive cluster covering algorithm (ACCO)

The other capable optimization technique, so-called Adaptive Cluster COvering (ACCO) algorithm was introduced by Solomatine (1999). The ACCO algorithm consists of four main principles: clustering, covering shrinking sub domains, adaptation and periodic randomization. Clustering is used in multistart algorithms to identify regions of attraction and to launch procedures of single-extremum search in each region. The idea of covering is used in set covering algorithms, and in pure direct random search. Adaptability update the algorithmic behavior depending on new information revealed about the problem under consideration. The last feature is periodic randomization. Due to the probabilistic nature of point generation, any strategy of randomized search may simply miss a promising region for search. In order to reduce this danger it is reasonable to re-randomize the initial population.

The ACCO algorithm is implemented based on the strategy of adaptive cluster covering that involves nine steps, as follows:

Step 1 (*initial sampling*) – Sample uniformly an initial population of N points in feasible domain X.

Step 2 (*initial reduction*) – Compute the function value f_i at each point and reduce the population by choosing p best points (with lowest f_i, objective function).

Step 3 (*initial clustering*) – Identify k_N clusters, such that the points inside a cluster are "close" to each other, and the clusters are "far" from each other. For each cluster, identify the smallest region (n-dimensional interval or a hull) for the subsequent search containing all points from the cluster. Set current region number $k = 1$. Set regional iteration number $e = 1$.

Step 4 (*start of subsequent regional iteration e*) – Sample r_k points inside region k, evaluate f at each of them, and choose s_k best points creating the set R_k. Reduce the region so that it includes the best points only.

Step 5 (*shifting to the center of attraction*) – Identify the 'center of attraction' of the region. This could be the best point or the centroid of the best subset. Shift the region so that its center coincides with the center of attraction.

Step 6 (*shrinking*). Reduce the size of the region so that its linear size would be v_k% of the previous one.

Step 7 (*stopping criteria for the current regional iteration*). Check criteria:
- C_1 is achieved when a fixed number e_l of regional iterations is reached;
- C_2 is achieved if average function value for the best u% points in R_k does not differ (fractionally) more than w from the same value computed at the previous iteration k-1 ;

- C_3 is achieved if during the last e_2 regional iterations; there was no improvement of the minimum estimate.

if (C_1 or C_2) and C_3, then begin

if $k = k_m$, then go to 8;

      prepare processing of the next cluster (region): set $k = k + 1$; $e = 1$;

end

else begin

      prepare the subsequent regional iteration: Check criterion C_{2B} (the average function value in the regional set R_k is larger than the average of all points evaluated so far, that is, the region seems to be 'non-interesting')

if C_{2B} then

      ('non-interesting' region) considerably decrease the sample size r_k for the next iteration by rd_k% (e.g., 30%)

else

      ('interesting' region) slightly decrease r_k by ri_k% (e.g., 5%);

      set $e = e + 1$.

end;

Go to 4.

Step 8 (*final accurate search*) – Construct the region with the linear size of q% of the domain interval around the best point found so far. Perform shifting and shrinking (steps 5 and 6) in a repetitive fashion until the stopping criterion C_1 or C_2 is satisfied.

Step 9 Check whether $l = T$. If yes, then STOP, otherwise set $l = l+1$ and go back to step 1.

8.3   Case Study

The research case study was concentrated on predicting the sea surges levels in two tidal stations located in the North Sea and in the Caribbean Sea.

The data used are located in the transition between December of 1994 and January of 1995. This period shows an important fluctuation of the surge level. Due to this characteristic this period is taken as testing data set when applying the predictive chaotic model. The data was split in training, testing and validation data set as described in Table 8-1. According with data-driven models it is possible to define three important data sets: training and cross-validation (testing). The training data set is defined as the predicted values in which the CM finds neighbors in the reconstructed phase space. Moreover, the cross-validation data set in this case is defined to apply the optimization methods in order to find an appropriate set of chaotic parameters. Finally the verification data set is defined to test the optimal parameter found in the optimization process.

TABLE 8-1: STATISTICAL DESCRIPTIONS OF TRAINING, TESTING AND VERIFICATION DATA SETS OF SURGES AT HOEK VAN HOLLAND TIDAL STATION.

Name	Data Set	Surges			
		Mean (cm)	Max. (cm)	Min. (cm)	Standard deviation
Hoek van Holland (HvH)	Training # points 43699 Rage for the Time indices: [1 – 43699] Range for the Date [1/1/1990 00:00:00 – 12/25/1994 19:00:00]	-0.96	220.33	-108.83	26.01
	Testing # points 300 Range for the Time indices: [43700 – 44000] Range for the Date [12/25/1994 20:00:00 – 1/13/1995 08:00:00]	19.37	157	-119	57.51
	Verification # points 300 Range for the time indices: [52420 – 52720] Range for the Date [12/25/1995 20:00:00 – 1/13/1996 08:00:00]	-7.81	82	-57	22.04

8.4   Model Setup

The predictive chaotic model (CM) to be optimized has the following characteristics. The CM uses a local cubic model in the reconstructed phase space. The number of neighbors is variable between 1 and k_{max}. There are two rules to filter the neighbors: the first one select the k_{max} closes neighbors from the current state and the second, from the last data set, select the neighbors which are between the minimum distance of the last set of point and twice this value. The method used to find neighbors in the phase space is Euclidean distance. The Taken's theorem is used to reconstruct the phase space. The prediction horizon is six hours.

8.4.1 Main experiment: predictive model for Hoek van Holland

8.4.1.1 Grid search

The values taken as reference to define the ranges for the time delay (τ) and embedding dimension (m) are taken from the nonlinear time series analysis. This analysis suggest a value of τ =11 (value obtained using mutual information) and m=6 (value obtained using false nearest neighbors). The typical parameters of a predictive chaotic model were defined: the local model used is cubic, the prediction horizon is 6hr and the neighbors are selected using the Euclidean distance. The data set was split as showed in Table 8-1. Finally different combinations of τ, m and number of maximum neighbors (k_{max}) were defined in order to explore the solution space by using a grid search. The values from 2 to 28 and from 2 to 28 were selected for τ and m, respectively. In addition, the values for the number of neighbors explored were: 5, 10, 15, 25, 50, 60, 80 and 100 (5832 combinations, time of simulation 24 days, one combination takes approximately 15min).

8.4.1.2 Randomized search

The grid search explored in the last section shows the general trend of the objective functions but does not cover the whole solution space. The main reason is due to the fact that it requires very intensive computation. The solution is done by means of randomized search techniques to explore the solution space. The algorithms used are: Genetic Algorithm (GA) and Adaptive Cluster Covering Optimization (ACCO).

The constant parameters used in the experiment are equal to the ones defined in the previous section for the grid search. We used the GA Tool Box in MATLAB. The configuration of the experiment can be divided in fourth important steps:

1) To formulate the fitness function which in this case is a function containing the predictive chaotic model (CM).
2) To define the function to perform the mutation (this function returns only integer values).
3) To determine the ranges for the chaotic parameters which are the decision variables (τ: time delay, m: embedding dimensions and k_{max}: number of maximum neighbors) in this particular case for the Hoek van Holland tidal station the ranges are: $\tau = [4\sim48]$, $m = [3\sim48]$ and $k_{max} =[1\sim100]$.
4) To define the parameters for the GA and those are: the number of generations (10) and population size (10). Finally the cross over fraction was defined as a high value in order to have diversity in each generation.

For ACCO we used the GLOBE software. The configuration of the problem in ACCO has three basic steps. The first one is to define the objective function that contains the CM. The second one defines the ranges for the decision variables as: $\tau = [4\sim48]$, $m = [3\sim48]$ and $k_{max} =[1\sim100]$. The third is to define the parameters of the ACCO algorithm (Solomatine 1999), including the number of clusters, number of iterations (all set to default values offered by software), and the initial population size (set to 10).

8.4.2 Additional experiment: predictive model for the San Juan station

For this experiment we also used an additional data set (water level at San Juan station) that became available at the time of this particular phase of work. The San Juan station is located in Puerto Rico. This station was taken in order to simulate the surge levels in a region affected by Hurricanes like the one presented in 2007 called Hurricane Felix. It was a destructive Category 5 hurricane on the Saffir-Simpson hurricane scale. In addition, it was the sixth named storm, second hurricane, and second Category 5 hurricane of the 2007 Atlantic hurricane season. The observed water level and predicted tide data were available and used for calculating the storm surge. The data has hourly continuous values from 2003 to 2008 (52608 hourly values). The water level was obtained from University of Hawaii Sea Level Center. The tides were predicted using a program called "wxtide32" (WXTide32, 2009). The water level is subtracted by tides for obtaining the surge levels (Table 8-2).

Furthermore, nonlinear time series analysis was carried out to find the proper values of time delay and embedding dimension by using mutual information and false nearest neighbors. The values found using this techniques are: $\tau = 4$ and $m=12$.

TABLE 8-2: STATISTICAL DESCRIPTIONS OF TRAINING, TESTING AND VERIFICATION DATA SETS OF SURGES AT SAN JUAN TIDAL STATION.

Name	Data Set	Surges			
		Mean (cm)	Max. (cm)	Min. (cm)	Standard deviation
San Juan (SJ)	Training # points 33599 Rage for the Time indices: [1 – 33599] Range for the Date [1/1/2003 00:00:00 – 10/31/2006 11:00:00]	105.22	132.7	78.9	7.67
	Testing # points 300 Rage for the Time indices: [33600-33900] Range for the Date [10/31/2006 12:00:00 – 11/13/2006 12:00:00]	112.72	130.50	98.90	6.67
	Verification # points 300 Rage for the Time indices: [42360-42660] Range for the Date [10/31/2007 12:00:00 – 11/13/2006 12:00:00]	109.38	125.20	94.90	6.63

8.4.2.1 Grid search

The estimation results from nonlinear time series analysis were used as a reference to define the ranges for the time delay (τ) and embedding dimension (m) for grid search optimization. The optimization suggested a value of $\tau = 4$ and $m=24$. The typical parameters of a predictive chaotic model were defined as in the previous case study. The data set was split as showed in Table 2. Finally different combinations of τ, m and number of maximum neighbors (k_{max}) were defined in order to explore the solution space by using a grid search.

Values from 2 to 28 and from 2 to 28 were selected for τ and m, respectively. Finally the values for the number of neighbors explored are: 5, 10, 15, 25, 50, 60, 80 and 100 (5832 combinations, time of simulation 24 days, one combination takes approximately 15min).

8.4.2.2 Randomized search

The configuration of GA and ACCO optimization was made the same as the one for the case study of Hoek van Holland tidal station.

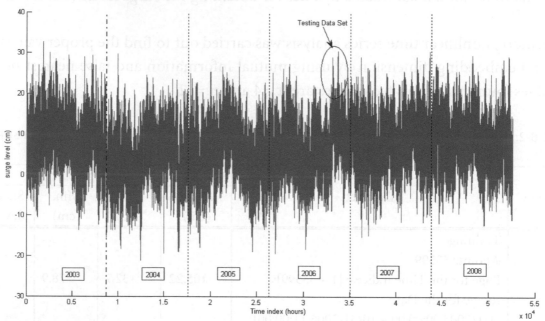

Figure 8-2: Surge time series for the San Juan tidal station between 2003 and 2008.

8.5   Model Results and Discussion

The comparison results of the predictive chaotic model optimization for Hoek van Holland and San Juan tidal stations are presented in Table 8-3 and Table 8-4, respectively. The comparison between observed an predicted values for the set of chaotic parameter found using optimization methods are presented in Figure 8-3, Figure 8-4, Figure 8-5 and Figure 8-6.

TABLE 8-3: OPTIMAL PARAMETER SETTING OF CHAOTIC MODEL FOR HOEK VAN HOLLAND OBTAINED FROM ACCO, GA AND GRID SEARCH (GS).

Method	τ	m	k_{max}	RMSE (cm)
GS	16	29	25	13.25
GA	24	20	26	13.75
ACCO	16	28	22	12.98

TABLE 8-4: OPTIMAL PARAMETER SETTING OF CHAOTIC MODEL FOR SAN JUAN OBTAINED FROM ACCO, GA AND GRID SEARCH (GS).

Method	τ	m	k_{max}	RMSE (cm)
GS	8	12	65	3.006
GA	8	12	55	3.007
ACCO	8	12	65	3.006

The results for the Hoek van Holland tidal station (Figures 5 and 6) shows that the optimal set of parameter reproduce the general tendency of the surge level for this location when using six hours for the prediction horizon. Moreover, the optimization results for San Juan tidal station (Figures 7 and 8) showed the same values of time delay and embedding dimension. This means that the time series under study have a sinusoidal tendency so when reconstructing the phase space the trajectories are smoother.

Based on these examples one may conclude that the randomized search and the grid search appear to be viable methods for identifying the appropriate embedding dimensions and time delay when predicting surge water levels. These methods can be combined to find appropriate values: first using a randomize algorithm to find local optimal points and then the search can be refined by using the grid search around those points. This solution is based, of course, on the ranges assigned to time delay (τ), embedding dimensions (m) and number of maximum neighbors (k_{max}), and these ranges for the time delay and embedding dimensions have to be obtained by chaotic analysis.

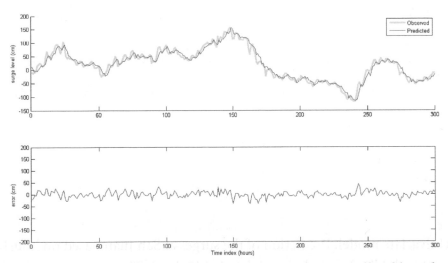

Figure 8-3: Univariate model. Prediction of the surges at Hoek van Holland based on hourly time series, testing data set (solid line) and observed data (wider line). Also the errors between observed and predicted values are shown $\tau=16$, $m=28$, $k=[1,22]$, RMSE=12.98 cm

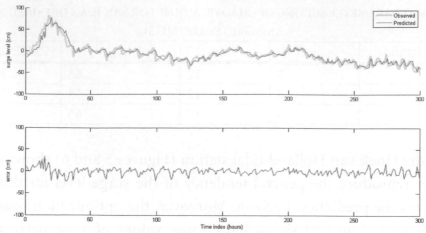

Figure 8-4: Univariate model. Prediction of the surges at Hoek van Holland based on hourly time series, validation data set (solid line) and observed data (wider line. The errors between observed and predicted values are shown with τ=16, m=28, k=[1,22], RMSE=7.08 cm.

The optimal values found using the randomized search are shown in Figure 5. It is possible to notice that the chaotic parameters found improve the performance of the predictive chaotic model. The predicted values near the peak are close to the measure ones. In Figure 6 the validation of the model under high surge and normal surge levels presents a considerably improvement in the prediction of the model. Also in the peak there is shift in the predicted values. This could be explain due the selection of the neighbors in the phase space is not totally accurate when building the local models.

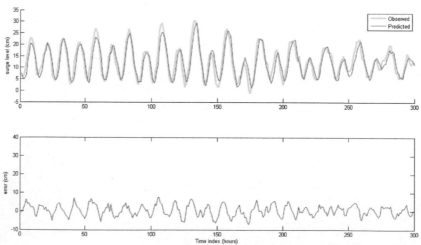

Figure 8-5: Univariate model. Prediction of the surges at San Juan based on hourly time series, testing data set (solid line) and observed data (wider line). The errors between observed and predicted values are shown (τ=8 m=12 k=[1,65]) with RMSE=3.01 cm.

The application of the optimization of predictive chaotic models in the Caribbean Ocean is presented. The results are shown in Figure 7 and 8. Figure 7 present the results for the prediction data set. The error presented is low in this case due the periodic values measure of surge levels. This sinusoidal behavior improves the performance of the predictive chaotic model. Furthermore, the error in the validation data set is also low an reflects that the optimization process find an appropriate set of chaotic parameters to prediction the surge levels.

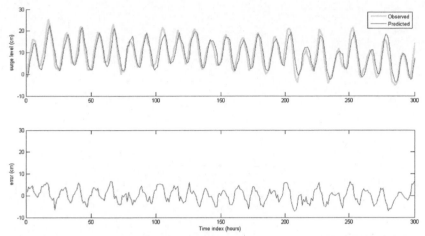

Figure 8-6: Univariate model. Prediction of the surges at San Juan based on hourly time series, validation data set (solid line) and observed data (wider line). The errors between observed and predicted values are shown ($\tau=8$, $m=12$, k=[1,65]) with RMSE=3.01 cm.

Finally this to examples shows that it is possible to improve the selection of the chaotic parameter by using optimization methods. This method can be used as a complementary to the well-known estimators in nonlinear time series analysis, such as correlation dimension and mutual information.

8.6   Summary

Chaotic analysis methods presented in Chapters 5 and 6 allow for finding the ranges of time delay and embedding dimension. In this Chapter we present the ways of finding the values of these parameters ensuring optmal performance of the predictive model. In application to storm surge prediction, exhaustive and randomized search algorithms appear to be viable methods for identifying the appropriate embedding dimension and time delay for the storm surge predictive model.   These methods can be also combined: first using a randomize algorithm to find local optimal points and then the search can be refined by using the exhaustive search around those points. The combination of methods reduces the computational effort when trying to find an optimal set of chaotic parameters.

The results show that the application of optimization methods improves the performance of predictive chaotic models for predicting storm surges. For the case study the optimal set of parameters when using randomized search are: $\tau=16$, $m=28$, $k_{max}=22$ and RMSE=12.98cm for the Hoek van Holland tidal station and $\tau=8$, $m=12$, $k_{max}=65$ and RMSE=3cm for the San Juan tidal station. These results are improved compared with the standard predictive chaotic model without optimization ($\tau=3$, $m=6$, $k=13$, no clustering) with RMSE=21.69. Overall the presented approach leads to more accurate chaotic predictive models.

CHAPTER 9:   REAL-TIME DATA ASSIMILATION USING NARX NEURAL NETWORK

*"We are the Borg. You will be assimilated. Your biological and
technological distinctiveness will be added to our own."*
Star trek: Voyager

This chapter presents a real-time data assimilation technique for storm surge predictive chaotic model using NARX neural network.

9.1   Introduction

Complexity of natural dynamical systems and sensitivity of corresponding models to initial conditions are the two major reasons of model prediction errors. The accuracy on defining initial condition is very crucial since the small error of the initial condition results in loss prediction ability of the model. Even if we had a perfect predictive model, all predictions start to diverge from the truth after certain time. However, these prediction errors can be corrected or reanalyzed once new observations available using data assimilation techniques. Data assimilation allows for making the best estimate of the necessary updates to the current model states or outputs so that the model can provide more reliable and accurate predictions.

A predictive model can be process-based or data-driven model which is seen to be composed of a set of equations (algorithms) that involve state variables and parameters. The model parameters typically remain constant while the state variables vary in time. Predictive model in real-time operation may take into consideration the new observation at the time of preparing for prediction. The feedback process of assimilating the new observation into the prediction procedure is so-called updating. The updating procedure can be classified into four different updating strategies based on the variables modified during the feedback process, as follows:

a) Updating input variables

Input uncertainties are often being the dominant source in prediction error. This method is often based on iterative procedures and the updating of input variables results in changing state variables.

b) Updating state variables

State variables can be adjusted for instance by updating state variable through simply substituting simulated variables by observed data from satellites. More comprehensive methodology using Kalman filtering theory can be used and/or integrated either with purely statistical transfer function models such as ARIMA.

c) Updating of model parameters

Adaptation of model parameters can be assessed through calibration using historical data. For a complex model, this adaption is extremely difficult because of a large number of parameters involved. In practice the use of this method is mostly confined to statistical black box models where it may be argued that no clear distinction exists between state variables and model parameters.

d) Updating of output variables (error prediction)

The difference error between model predictions and observation (model error) are usually found to be serially correlated. This allows for predicting the future values of these errors by means of time series models such as ARMA model. The model predictions can be improved by adding the error predictions from the ARMA error model. This method which is often referred to as error prediction is the most widely used in practice (WMO, 1992). Earlier studies on model error prediction have been conducted by Jamieson *et al.* (1972), Lundberg (1982) and Szöllösy-Nagy *et al.* (1983).

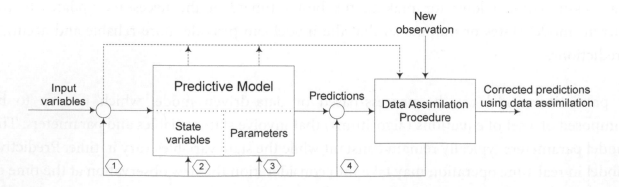

Figure 9-1: Schematic diagram of predictive model with data assimilation on different updating strategies (modified WMO, 1992).

The data assimilation techniques have been broadly studied for more than a decade. A number of different data assimilation techniques have been developed (Heemink *et al.*, 1997; Bouttier & Courtier, 1999; El Serafy *et al.*, 2007; Evensen, 2007). Heemink et al. (1997), Verlaan & Heemink (1997) and Robaczewska et al. (1997) have developed the data assimilation techniques for storm surge numerical models particularly in the Netherlands.

For improving the accuracy and predictability of chaotic storm surge models, this research focuses on the use of artificial neural networks (ANNs) for the task of data assimilation. ANNs have been successfully applied to a number of time series prediction and modeling tasks in various fields of sciences and engineering (Haykin, 1999; Bishop, 2006)(Haykin, 2008). In particular, when the time series is noisy and the underlying dynamical system is nonlinear, ANN models frequently outperform standard linear techniques, such as the Box–Jenkins models (Box *et al.*, 1994). Some experiments and research on the applications of ANNs for the task of data assimilation have been explored. Härter and de Campos Velho (2008) employed radial-basis function neural network for data assimilation of 1D shallow water model by emulating Ensemble Kalman filter. Furthermore, a multi-layered perceptron neural network was utilized to imitate particle filter for data-assimilation of Lorenz system (Furtado *et al.*, 2008).

In this work, the Nonlinear AutoRegressive eXogenous inputs (NARX) neural network is used for data-assimilating a chaotic storm surge model with new observations. The NARX neural network is a recurrent neural network model with the limited feedback - only from the output neurons and not from the hidden ones and has exogenous and endogenous time-delayed inputs. In practice, NARX neural network has several advantages. It has been reported that the NARX neural network is more powerful than conventional recurrent networks (Horne & Giles, 1995). The gradient-descent learning can be more effective in NARX neural network due to its embedded memory which provides a shorter path for propagating gradient information when the network is unfolded in time to back-propagate the error signal. Thus, it can reduce the network's sensitivity to the problem of the long-term dependencies (Menezes & Barreto, 2008). The application of self-organizing NARX neural network in the role of predictor for a chaotic time series was reported by Barreto and Araujo (2001). They verified that NARX neural network is more powerful than conventional recurrent networks and can reduce the network's sensitivity to the problem of the long-term dependencies. However, we could not find references to research specifically where NARX neural network is used as the data-assimilation device with a predictive chaotic model is employed.

For the purpose of data assimilation, a NARX neural network is trained using the historical prediction errors and observations. Subsequently, the trained NARX neural network can correct the model predictions with the newly available observations fed as the exogenous inputs. The NARX data assimilation is performed in nearly real-time process since it does not require a lot of computation in comparison to variational or sequential methods. This proposed method was implemented to data-assimilate chaotic storm surge models for the North Sea.

9.2 NARX Neural Network

9.2.1 Network Architecture

The Nonlinear AutoRegressive eXogenious inputs (NARX) neural network is a class of discrete-time recurrent neural networks that can be mathematically represented as:

$$y(n+1) = f[y(n), \dots, y(n-d_y+1); x(n-k), x(n-k+1), \dots, x(n-d_x-k+1)] \qquad (9.1)$$

where $x(n) \in \Re$ and $y(n) \in \Re$ denote, respectively, the input and output of the model at discrete time step n, while $d_x \geq 1$ and $d_y \geq 1$, $d_y \geq d_x$, are the input-memory and output-memory orders, respectively. The parameter k is a delay term, known as the process dead-time (Haykin, 1999). In general, we assume $k=0$, thus obtaining the following NARX model:

$$y(n+1) = f[y(n), \dots, y(n-d_y+1); \ x(n), x(n-1), \dots, x(n-d_x+1)] \qquad (9.2)$$

which may be written in vector form as:

$$y(n+1) = f[\boldsymbol{y}(n); \boldsymbol{x}(n)] \qquad (9.3)$$

where the vectors $\boldsymbol{y}(n)$ and $\boldsymbol{x}(n)$ denote the output and input regressors, respectively. The nonlinear mapping $f(.)$ is generally unknown and can be approximated, for example, by a standard MLP network. The resulting connectionist architecture is then called a NARX neural network, a powerful class of dynamical models which has been shown to be computationally equivalent to Turing machines (Siegelmann *et al.*, 1997). Figure 9-2 shows the architecture of NARX neural network with three hidden neurons, d_x delayed inputs and d_y delayed outputs.

9.2.2 Learning Algorithm

Error surfaces of dynamic networks can be more complex than those of static networks. Learning of such dynamic network like NARX is more likely to be trapped in local minima. A dynamic back-propagation (BP) algorithm is required to compute the gradients, which is computationally more intensive than that for a static BP.

One of the intensively used learning algorithms is Levenberg-Marquardt optimization algorithm. It combines the advantages of the simple gradient descent and the Newton's method algorithm, and has rapid convergence and robust performance In this paper, we use a modified Levenberg-Marquardt optimization algorithm with the inclusion of Bayesian regularization technique (Lin *et al.*, 1997) for training the NARX neural network. It minimizes a combination of squared errors and weights to produce a NARX model which generalizes well.

Regularization is a way of dealing with the negative effect of large weights which cause excessive variance of outputs. The idea of regularization is to make the network response smoother through the modification in the objective function by adding penalty term consisting of the squares of all network weights. This additional term favors small values of weights and decreases the tendency of a model to overfit noise in training data set. Mackay (2003) introduced a technique, so-called Bayesian regularization which automatically sets the optimal function to get the best generalization based on Bayesian inference method. The Bayesian optimization of the regularization parameters requires the computation of Hessian matrix at the minimum point. This can be done using Gauss-Newton approximation to Hessian matrix if Levenberg-Marquardt optimization algorithm is used to locate the minimum point.

9.3   NARX Data Assimilation

A process of approximating the true state of a physical system at a given time is called analysis. The information on which the analysis is based includes the observational data and the model of the physical system, together with some background information on initial and boundary conditions and, possibly, the additional constraints on the analysis. An analysis can be very simple, for example a spatial interpolation of observations. However, much better results can be obtained by involving the dynamic evolution of the physical system into the analysis. Data assimilation combines time distributed observations and a dynamic model. It aims at accurate re-analysis, estimation and prediction of an unknown, true state by merging observed information into a model (Evensen, 2007).

A particular challenge in predicting the time-evolution of a dynamical system is the nonlinearity of such system and the corresponding sensitivity to initial conditions. It is well known that even if we had a perfect prediction model, all predictions start to diverge from the truth after a finite time. These diverged or error predictions can be corrected or re-analyzed once new observations available via data assimilation. Variational analysis (3D/4DVar) and sequential method (i.e. Ensemble Kalman filter) are two data assimilation methods which are commonly used in physically-based (numerical) models (Heemink & Metzelaar, 1995; Kalnay, 2003). The variational data assimilation consists of minimizing a predefined cost function that measures the difference between model predictions and observations over a certain time interval. In the sequential data assimilation, a recursive updating of the model solution is made during a forward integration where model predictions and observations are weighted according to the associated uncertainties. In general, data assimilation process can be mathematically represented by:

$$\mathbf{x}^a = \mathbf{x}^f - \mathbf{W}[\mathbf{y}^{obs} - \mathbf{H}(\mathbf{x}^f)] \tag{9.4}$$

where $\mathbf{x}^a$ is the values of the analysis, $\mathbf{x}^f$ is the model prediction (also known as beckgound field), $\mathbf{W}$ is the weight (covariance) matrix as a distance function between predictions and observations, $\mathbf{y}^{obs}$ is the observations, $\mathbf{H}$ is the observation system and $[\mathbf{y}^{obs} - \mathbf{H}(\mathbf{x}^f)]$ is the innovation. Many data assimilation methods are based on this equation (Eq.8), but differ by the approach to combine the model predictions and observations to produce the analysis, numerical cost function and optimality.

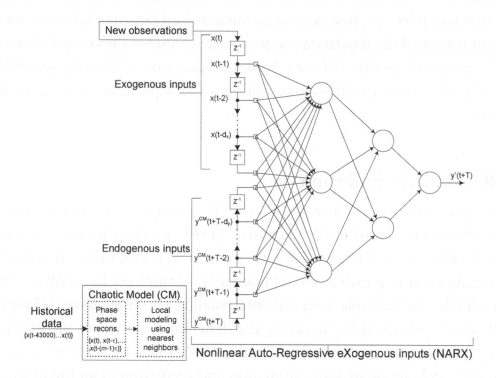

Figure 9-2: Predictive chaotic model with data assimilation using NARX neural network. The predictive chaotic model predictions (using phase space reconstruction and local modeling) and new observations are fed into endogenous and exogenous inputs of NARX neural network networks, respectively.

A data assimilation technique using NARX neural network is introduced for combining predictive chaotic model predictions and observations (as schematized in Figure 9-2) for improving the accuracy of predictive chaotic model predictions. The predictions (output) of predictive chaotic model with time-delayed inputs (TDI) and new observations are fed as endogenous and exogenous tapped delay inputs (TDL), respectively, in the NARX neural network. Parallel NARX architecture is employed here since only one-step prediction or assimilation process is required. The following sections describe the case study and the implementation and experimental results of this data assimilation technique for assimilating the predictive chaotic models of storm surge dynamics in the North Sea.

9.4   Data Description

The data used comes from the same location (Hoek van Holland) as described in the previous experiments, but it was measured at different periods of time; one of the reasons for this was to be able to include a particularly interesting extreme storm surge in the North Sea on November 9th, 2007. The physically-based models of several European institutions provide the predictions of storm surges. Each time series begins at 00:00 January 1st, 2003 and is available until 23:50 December 31st, 2007 with some missing values. This results in 262944 continuous samples in total for the 10 min data, and 43824 for the hourly data. Due to different resolutions of European storm surge model time steps, the hourly time series data used for further analysis and model prediction were made by an averaging technique.

TABLE 9-1: DATA SEPARATION OF HOURLY SURGE TIME SERIES FOR TRAINING, VALIDATION AND VERIFICATION DATA SETS (AT HOEK VAN HOLLAND).

Date	Surge Data Sets (hourly)				
	Training	Validation		Verification	
		Non-stormy	Stormy	Non-stormy	Stormy
Start	11Sep05	01Jun04	08Oct04	1Sep07	15Oct07
End	31Aug07	31Aug04	31Dec04	14Oct07	20Nov07

Table 9-1 shows the data separation of hourly surge time series at HvH for training, validation and testing data sets. The validation and testing data sets are split into non-stormy and stormy periods with the objective to investigate the model performances during normal and extreme storm surge conditions.

9.5   Model Results and Discussion

9.5.1 Estimating delay time and embedding dimension

Some methods based on the theory of nonlinear dynamics and deterministic chaos (Hegger et al., 1999; Siek & Solomatine, 2010a) were used to analyze surge time series for HvH tidal station and to estimate the proper values of m and τ. The appropriate delay time τ for the reconstruction of the surge dynamics at HvH was assessed using the methods of autocorrelation function and the first minimum average mutual information (Fraser & Swinney, 1986), as presented in Figure 9-3. The first minimum mutual information is $\tau=10$ hours.

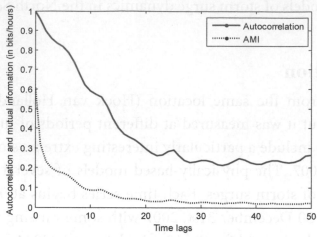

Figure 9-3: The autocorrelation function (solid line) and mutual information (dashed line) as a function of time lags for the hourly surge time series at Hoek van Holland location.

For estimating embedding dimension m, the correlation dimension d_c, which is based on the correlation integral or function analysis (Grassberger & Procaccia, 1983a) was utilized. Obtaining a non-integer, finite d_c for a time series demonstrates fractal scaling and indicates possible chaotic behavior. Figure 9-4 shows that the correlation exponent increases with an increase of the embedded dimension up to a certain value and further saturates. A nonlinear Gaussian noise reduction was applied to obtain a better estimation of correlation dimension. The saturation value of the correlation exponent is 7.8, which indicates the presence of an attractor in the surge dynamics. Taking into account the estimation of the embedding dimension m, if one uses the Taken's embedding theorem the embedded dimension (integer number) of the manifold which contains the attractor is $m=2d_c+1 \approx 17$.

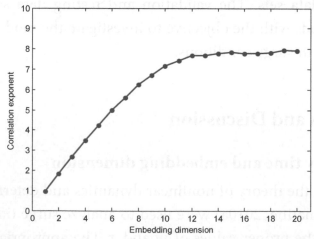

Figure 9-4: Relationship between the correlation exponent ν and embedding dimension m.

Another method for estimating the proper values of m is false nearest neighbors (FNN) method, which measures the percentages of false nearest neighbors between successive embedding dimensions. The FNN result also gives an estimation of $m=17$. Figure 9-5 shows that the percentage of the FNN drops to about 30% with $m=17$, and remains unchanged for further increase of m. However, we investigated further on the optimal choice of m using an exhaustive search optimization with respect to the model performance over cross-validation data sets during non-stormy and stormy periods. The result showed the optimal value of m is 18 and this value was used for building the predictive chaotic model and prediction.

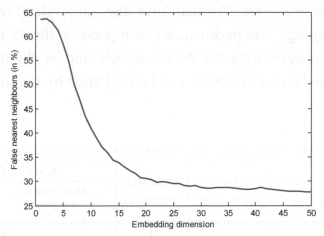

Figure 9-5: Percentage of the false nearest neighbors as function of the embedding dimension m.

In addition, the stability and predictability of the surge dynamics was investigated by the use of Lyapunov exponents (Sano and Sawada, 1985). Figure 9-6 depicts the Lyapunov spectrum with the largest Lyapunov exponent estimated as $\lambda_1=0.07$. This indicates a loss of information of 0.07 bits/hour during the dynamical evolution of the system, and thus loss of predictive capabilities. The Lyapunov spectrum contains a large negative exponent $\lambda_{18}=-0.7$ which indicates presence of strong dissipation mechanisms in the surge dynamics. The presence of positive Lyapunov exponents and the fact that $\Sigma\lambda_i=-1.56<0$, provide strong evidence that such system is driven by deterministic chaos.

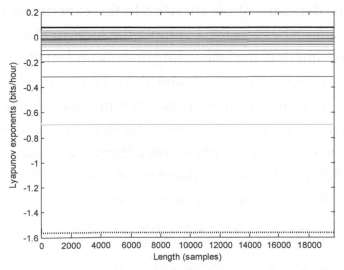

Figure 9-6: Lyapunov spectrum for the hourly surge time series at Hoek van Holland tidal station for $m=18$ dimension. The largest Lyapunov exponent (bold black line) is positive and a sum of global Lyapunov exponents (dashed line) is negative.

9.5.2 European operational storm surge models

Table 9-2 lists the performances of the operational physically-based numerical storm surge models from European institutions, namely: BSH Germany, DMI Denmark, DNMI Norway, KNMI Netherlands and MUMM Belgium. These models have diversity on prediction horizons, time steps, data availability and data assimilation frequencies. For comparison, these data were transformed into the same data resolution (hourly) by interpolating and averaging. The performance comparison shows that BSH storm surge model from Germany outperforms the other models and is slightly better than KNMI model from Netherlands (such differences can be explained by the difference in periodicity of data assimilation).

TABLE 9-2: PERFORMANCE COMPARISON BETWEEN SEVERAL EUROPEAN OPERATIONAL STORM SURGE MODELS.

Models	Forecast horizon and data assimilation (hours)	Time step (minutes)	RMS Error (in cm)	
			Non-stormy Period	Stormy Period
BSH	12	10	7.96	10.92
DMI	6	10	9.97	12.58
DNMI	12	60	10.82	12.35
KNMI	48(6)*	10	8.95	**11.62**
MUMM	6	10	15.42	19.15

*Data assimilation done every 6 hours, for other models unknown

9.5.3 Chaotic storm surge models

Adaptive local models (constant, linear, quadratic and 3rd- order polynomial) based on dynamical neighbors were implemented and used in the reconstructed phase space of the surge at HvH to map the dynamics of the attractor. In this experiment only the information from the surge hourly time series was used to build the local models. The sensitivity of the choice of the local approximation, the embedding dimension (m), the time delay (τ) and the number of neighbors (k) were investigated, and these values were optimized. The results showed that the use of 3rd-order polynomial function provides the best predictive chaotic model performance. The surge predictions were further compared with neural network (MLP) models. The same reconstructed phase space (input data) was used to train different MLP NNs, using different structures (number of hidden layer/nodes and transfer functions). The proper phase reconstruction is as follows:

$$\mathbf{Y_t} = \{s_t^{hvh}, s_{t-10}^{hvh}, s_{t-20}^{hvh}, ..., s_{t-180}^{hvh}\} \tag{9.5}$$

Table 9-3 summarizes the performance of the predictive chaotic models for the surge predictions and the results of the best performing NN for prediction horizons of 1, 6, 12, 24 and 48 hours. The results indicate that predictive chaotic models outperform MLPs significantly for both non-stormy and stormy periods. In Figure 9-7, the amplitudes of the extreme positive surges which is actually the storm surge on November 9th, 2007 are not correctly predicted by NN. The predictive chaotic model is able to predict this extreme storm surge better, but still cannot reach the surge peak. This is due to the fact that for the rare peaks the predictive chaotic model does not have 'true' neighbors (no such high surges in the past) and lack of extrapolation ability of the built local models corresponding to this extreme surge.

TABLE 9-3: PERFORMANCE COMPARISON OF ARTIFICIAL NEURAL NETWORK AND PREDICTIVE CHAOTIC MODEL (WITHOUT METEOROLOGICAL PREDICTIONS).

Models	RMS Errors (in cm) for Different Prediction Horizons				
	1hr	6hr	12hr	24hr	48hr
Non-stormy Period - m=18, τ=10, k=[50 100]					
No. hidden nodes	6	11	9	1	3
Neural networks	6.16	11.65	15.96	20.44	19.61
Predictive chaotic model	6.56	12.46	16.63	22.50	21.77
Stormy Period - m=18, τ=12, k=[9 100]					
No. hidden nodes	9	14	8	11	14
Neural networks	8.24	16.98	27.32	33.66	**33.74**
Predictive chaotic model	7.80	19.60	30.31	37.11	**36.35**

9.5.4 Data assimilation using NARX neural network

The numbers of tapped input delays and hidden nodes of NARX were optimized by an exhaustive search procedure. The frequencies of data assimilation for the built predictive chaotic models were set to be 6, 12 and 24 hours. This data assimilation frequency was made with considerations to tidal cycle and the operational data assimilation of the European storm surge model, like KNMI (Dutch meteo service), is done for every 6 hours. The data assimilation process was performed in nearly real-time since the process is one-step prediction of the trained NARX neural network given some new observations and recent chaotic model predictions.

Table 9-4 lists 48-hours ahead predictive chaotic model prediction errors with data assimilation for non-stormy and stormy periods. It is shown that the predictive chaotic model with data assimilation has a significant increase of prediction accuracy than the one without data assimilation and the European physically-based models. In 48-hours prediction during stormy period, the predictive chaotic models with (6 hours) and without data assimilation and KNMI model (with EnKf data assimilation) have RMS errors of 5.54cm, 36.35cm and 11.62cm, respectively (see bold numbers in Table 9-2, Table 9-3 and

Table 9-4). This demonstrates that the incorporation of data assimilation method in predictive chaotic model can increase the accuracy of predictions and extend the predictability of such model which is typically only reliable for short-term prediction.

TABLE 9-4: PERFORMANCES OF 48 HOURS CHAOTIC MODEL PREDICTIONS WITH DIFFERENT FREQUENCIES OF NARX NETWORK DATA ASSIMILATION.

Predictive chaotic model with Different Frequency of Data Assimilation Using NARX (RMS errors in cm)			
	6hr	12hr	24hr
Non-stormy period			
No. tapped input delays	20	20	20
No. hidden nodes	20	20	20
RMSE	2.05	1.93	2.08
Stormy period			
No. tapped input delays	20	20	19
No. hidden nodes	17	20	14
RMSE	**5.57**	15.87	20.09

9.6   Summary

A real-time data assimilation technique using NARX neural network is introduced and implemented to re-analyze and improve the predictive chaotic model predictions of storm surge in the North Sea. The predictive chaotic model with data assimilation has demonstrated a pronounced capability for reliable and accurate prediction outperforming standard predictive chaotic model, ANN (MLP) model and the European numerical storm surge models. For 48-hours ahead prediction at Hoek van Holland station during stormy period, the predictive chaotic models with NARX data assimilation (every 6 hours) outperforms the standard predictive chaotic model by 553% and the KNMI numerical model with EnKf data assimilation by 109%. This demonstrates the effectiveness of the proposed method.

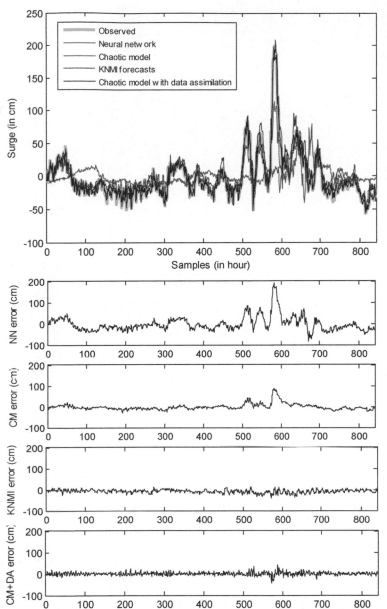

Figure 9-7: Storm surge predictions of the artificial neural network (RMSE=33.74cm), predictive chaotic model (RMSE=36.35cm), operational KNMI model with EnKf data assimilation (RMSE=11.62cm) and predictive chaotic model with NARX data assimilation in every 6 hours (RMSE=5.57cm) at Hoek van Holland station for the stormy period (15-Oct-2007 till 20-Nov-2007). The prediction horizon is 48 hours. The four bottom figures show the errors.

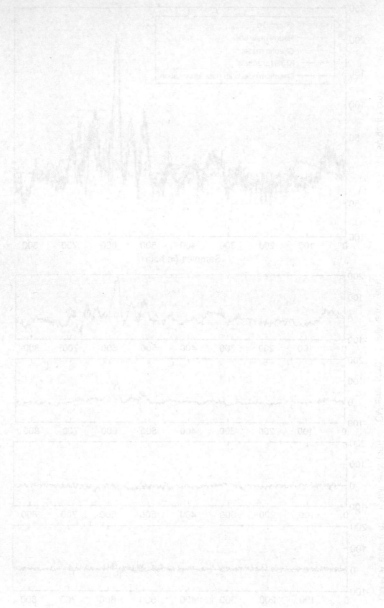

CHAPTER 10: ENSEMBLE MODEL PREDICTION

"United we stand, divided we fall."

Aesop

This chapter introduces novel ensemble techniques, dynamic averaging and dynamic neural networks, and their use to build ensembles of chaotic storm surge models in high dimensional space.

10.1 Introduction

Ensemble model predictions have been viewed for some decades as an effective way to improve the prediction performance over what the individual models can provide. It is often worthwhile to seek a combination of several prediction models rather than to select only the best one among them, which might be only marginally the best. Therefore, ensemble models now become the main topic in widespread use of model prediction in many fields, (e.g., in hydrometeorology and geosciences). Some recent researches on the use of prediction combination model for meteorological prediction and time series prediction have been conducted by Palmer *et al.* (2005), Wichard & Ogorzalek (2004) and Zhu (2005). Research efforts are now aimed at determining what kind of predictions benefit most from such combinations, and what combination techniques are optimal in a various situations.

10.2 Principles of Ensemble Model Prediction

There are various models for storm surge prediction for the North Sea nowadays. However, when building a prediction model, it is not an easy task to choose a reliable model, since on one hand, no model is powerful and general enough to outperform the others for all types of circumstances; on the other hand, every model has some degree of uncertainty. Instead of using a single model, alternatively the predictions from various models are combined in such a way that a more reliable and accurate prediction can be obtained.

Numerical models and data assimilation systems have improved enormously over recent years so that today's 3-day prediction can be as good as a 1-day prediction 20 years ago. Despite this, a prediction system looking a few days ahead can frequently be quite wrong, and even 1-day predictions can occasionally have large errors. The reason for this lays in the chaotic nature of the system, which means that very small errors in the initial conditions can lead to large errors in the prediction, the so-called butterfly effect. This means that a perfect prediction system cannot be found because every detail of the initial state of the system is never well observed. Tiny errors in the initial state will be amplified such that after a period of time the prediction becomes useless. This sensitivity varies from time to time. The uncertainties in the predictions can become large as the prediction horizon is larger (Lorenz, 1963; Kalnay, 2003).

To cope with this uncertainty, an ensemble prediction is used. Instead of running just a single prediction, the model is run a number of times from slightly different starting conditions. The complete set of predictions is referred to as the ensemble and individual predictions within it as ensemble members. The initial differences between ensemble members are very small so that if we compared members with observations it would be impossible to say which members fitted the observations better. All members are therefore equally likely to be correct, but when we look longer time ahead the predictions can be quite different. On longer term the members can differ radically and then more caution is required.

There are two categories of approaches in combining predictions. The first one is the ensemble approach, by which a set of predictions are produced on the same task with different models (or one model with different inputs), and then the predictions are combined. The second one is the modular approach, under which a task or problem is divided into a number of subtasks (regimes), and the complete task solution requires the contribution of all of the individual regimes.

10.2.1 Information-theoretic model selection

Most of information scientists do not believe in the notion of true models. Models, by definition, are only approximations to unknown reality or truth; there are no true models that perfectly reflect full reality. Furthermore, the best model for analysis of data depends on sample size; smaller effects can often only be revealed as sample size increases. Hence, a given set of data has only a finite amount of information. The unachievable objective of model selection is to find a perfect translation such that no information is lost in going from the data to a model of the information in the data. However, one can attempt to find a model of the data that is best in the sense that the model loses as little information as possible (Burnham & Anderson, 2002). This leads to Kullback-Leibler (K-L) information $I(f,g)$ that was formulated based on Boltzmann's concept of entropy. It measures the information loss when model g is used to approximate full reality f. The K-L distance between conceptual truth f and model g is defined for continuous functions as the integral of:

$$I(f,g) = \int f(x) \log\left(\frac{f(x)}{g(x\,|\,\theta)}\right) dx \qquad (10.1)$$

where $I(f,g)$ is the information loss when model g is used to approximate truth f, log denotes the natural logarithm and f and g are n-dimensional probability distributions. The main goal is to seek an approximating model that loses as little information as possible; this is equivalent to minimizing $I(f,g)$ over the models in the set. The other information criteria for finding the best model are Akaike's information criterion (AIC) and Takeuchi's information criterion (TIC) (Akaike, 1974; Takeuchi, 1976).

Another possibility is to combine the entire set of models by using model averaging. Model averaging computes a weighted estimate of the predicted value. Akaike weights can be used for weighting predictions. Thus, if a parameter θ is in common over all models (as θ_i in model g_i), then the weighting average is:

$$\hat{\bar{\theta}} = \sum_{i=1}^{R} w_i \hat{\theta}_i \qquad (10.2)$$

where $\hat{\bar{\theta}}$ denotes a model averaged estimate of θ. This approach has practical advantage that has better precision and reduced bias compared to the approach of selecting one best model.

10.2.2 Bayesian model averaging

Bayesian approach to statistics considers the problem of estimating some probability (such as a future outcome or a noisy measurement), based on measurements of our data, a model for these measurements, and some model for our prior beliefs about the system. Let us consider a standard two-stage model, where we write our data measurements as a vector $\mathbf{y}=[y_1, y_2,...,y_n]$, and our prior beliefs as some vector of random unknowns θ. The model of measurements can be written as a conditional probability distribution (or likelihood) $p(y|\theta)$, and also the prior as $p(\theta|\eta)$, where η is some hyper-parameter.

Typical statistical analysis, such as regression analysis, typically proceeds conditionally on one assumed statistical model. Often this model has been selected from among several possible competing models for the data, and the data analyst is not sure that it is the best one. Other plausible models could give different answers to the scientific question at hand. This is a source of uncertainty in drawing conclusions, and the typical approach, that of conditioning on a single model deemed to be "best", ignores this source of uncertainty, thus underestimating uncertainty.

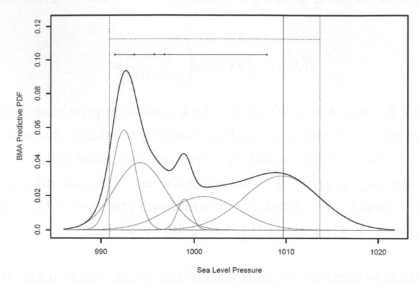

Figure 10-1: BMA predictive PDF (thick curve) and its five components (thin curves), the ensemble member predictions and range (solid horizontal line and bullets), the BMA 90% prediction interval (dotted lines), and the verifying observation (solid vertical line) (Raftery *et al.*, 2005).

Bayesian model averaging overcomes this problem by conditioning, not on a single "best" model, but on the entire ensemble of statistical models first considered (Raftery *et al.*, 2005). In the case of a quantity y to be prediction on the basis of training data y^T using K statistical models [$M1,...,M_K$], the law of total probability tells us that the prediction PDF, $p(y)$, is given by:

$$p(y) = \sum_{k=1}^{K} p(y \mid M_k) p(M_k \mid y^T) \tag{10.3}$$

where $p(y|M_k)$ is the prediction PDF based on model M_k alone, and $p(M_k|y^T)$ is the posterior probability of model M_k being correct given the training data, and reflects how well model M_k fits the training data (Figure 10-1). The posterior model probabilities add up to one, so that $\sum_{k=1}^{K} p(M_k \mid y^T)=1$, and they can thus be viewed as weights. The BMA PDF is a weighted average of the PDFs given the individual models, weighted by their posterior model probabilities. BMA possesses a range of theoretical optimality properties and has shown good performance in a variety of simulated and real data situations.

We now extend BMA from statistical models to dynamical models. The basic idea is that for any given prediction there is a "best" model, but we do not know what it is, and our uncertainty about the best model is quantified by BMA. Once again, we denote by y the quantity to be predicted. Each deterministic prediction, f_k, can be bias-corrected, yielding a bias-corrected prediction $\tilde{f}_k$. The prediction f_k is then associated with a conditional PDF, $g_k(y \mid \tilde{f}_k)$, which can be interpreted as the conditional PDF of y conditional on $\tilde{f}_k$, given that f_k is the best prediction in the ensemble. The BMA predictive model is then:

$$p(y \mid f_1,...,f_K) = \sum_{k=1}^{K} w_k g_k(y \mid f_k) \qquad (10.4)$$

where w_k is the posterior probability of prediction k being the best one, and is based on prediction k's skill in the training period. The w_k's are probabilities and so they add up to 1, i.e. $\sum_{k=1}^{K} w_k = 1$. How to estimate w_k is described as the following.

When predicting sea level and surge, it often seems reasonable to approximate the conditional PDF by a normal distribution centered at $\tilde{f}_k$, so that $g_k(y \mid \tilde{f}_k)$ is a normal PDF with mean $\tilde{f}_k$ and an ensemble-member-specific standard deviation, σ_k. We denote this situation by:

$$y \mid \tilde{f}_k \sim N(\tilde{f}_k, \sigma_k^2) \qquad (10.5)$$

and we will describe how to estimate σ_k^2 in the next subsection. In that case, the BMA predictive mean is just the conditional expectation of y given the predictions, namely:

$$E[y \mid f_1,...,f_K] = \sum_{k=1}^{K} w_k \tilde{f}_k \qquad (10.6)$$

This can be viewed as a deterministic prediction in its own right, and can be compared with the individual predictions in the ensemble or the ensemble mean. Subsequently, maximum likelihood of model parameters, w_k and σ_k^2, are estimated by the EM (expectation-maximization) algorithm. Figure 10-2 shows an example of relative frequency of the ensemble prediction members before (under dispersive) and after calibration (equally probable) using BMA.

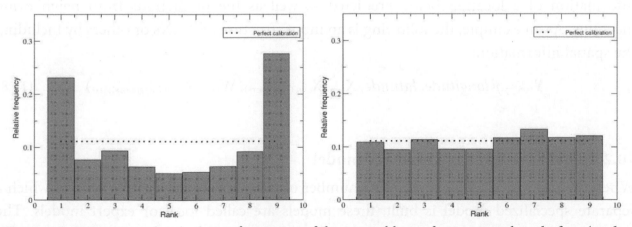

Figure 10-2: An example of relative frequency of the ensemble prediction members before (under dispersive) and after calibration (equally probable) using BMA.

10.2.3 Ensembles with spatial information

Kriging is a spatial estimation of the local continuous function using linear models. The term "kriging" was introduced by G. Matheron in 1963 after the name of D.G. Krige, who was a mining engineer in South Africa. The method interpolates the value $Z(x_0)$ of a random field $Z(x)$ at an unobserved location x_0 from observations of the random field at nearby locations. Kriging computes the best linear unbiased estimator of $Z(x_0)$ based on a stochastic model of the spatial dependence quantified either by the variogram $\gamma(x,y)$ or by expectation $\mu(x)=E[Z(x)]$ and the covariance function $c(x,y)$ of the random field. The kriging estimator is given by a linear combination:

$$\hat{Z}(x_0) = \sum_{i=1}^{n} w_i(x_0)Z(x_i) \tag{10.7}$$

of the observed values $z_i = Z(x_i)$ with weights $w_i(x_0)$, $i=1,...,n$ chosen such that the kriging variance or error minimized subject to the unbiasness condition. Several types of kriging are: simple kriging, ordinary kriging (with trend), universal kriging (using general linear trend model), IRFk-kriging (using polynomial function), indicator kriging (using indicator functions to estimate transition probabilities, multiple indicator kriging (working with a family of indicators), disjunctive kriging (a nonlinear generalisation of kriging), lognormal kriging (interpolates by means of logarithms, Bayesian kriging and so forth. Further information about kriging techniques can be found in Kanevski & Maignan (2004). The kriging family methods can be used for ensemble model predictions with inclusion of spatial information.

Moreover, the use of machine learning techniques, like ANNs, Fuzzy rule based system, predictive chaotic models, Bayesian model averaging can be explored and investigated for combining model predictions with inclusion of spatial information. The inclusion of spatial information into modeling can be done by inserting the spatial information as inputs of the machine learning models. The spatial information in this case can be the geospatial information of a location being predicted as well as the predictions from neighboring locations. As an example, the following is an input structure of ANNs or others by including the spatial information:

$$\mathbf{Y}_{t+n} = f(longitude, latitude, \mathrm{X}_{t+n}, \mathrm{X}_{t+n\,(neighbors)}, \mathrm{W}_{t+n}, \mathrm{W}_{t+n\,(neighbors)}, ...) \tag{10.8}$$

10.2.4 Machine learning: modular model

When the input space is divided into a number of subspaces or regions for each of which a separate specialized model is built, these models are called local or expert models. The resulting model is called a modular model (MM). A model consisting of multiple models whose outputs are combined is often called a committee machine (CM) (Haykin, 1999). The CM can be classified into (Solomatine & Siek, 2006):

- Hard splits (modular local models): input data is split and outputs are combined
- Soft splits (mixture of experts, boosting, bagging)
- No splits (ensembles): models are trained on whole input data set and the outputs are combined using a weighting scheme

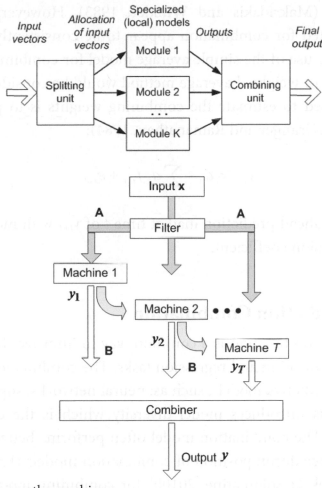

A: distribute data between the machines

B: pass the statistically sampled data (each using different distribution) to the subsequent machines

Figure 10-3: Modular models: input data is split and fed into multiple models whose outputs are combined (Shrestha & Solomatine, 2006; Solomatine & Siek, 2006).

10.3 Linear Prediction Combination

Essentially, the model combination method is a weighted average of the outputs of combination members. The combination prediction technique is normally used to provide probabilistic predictions. There are two main issues about this approach. First, how to select a set of models and generate an combination of predictions to be combined, and second, how to estimate the combining weights so as to minimize the out-of-sample prediction errors.

As for the estimation of combining weights, some studies show that equally weighted combination, namely, the simple average method (SAM), can produce predictions that are better than those of the individual models, and its accuracy depends mainly on the number of the models involved and on the actual prediction ability of the specific models included in the simple average (Makridakis and Winkler, 1983). However, when some of the individual models selected for combination appear to be consistently more accurate than others, in which case the use of the simple average model for combining predictions can be quite inefficient, the use of weighted average method would be considered. One of the most common procedures used to estimate the combining weights is to perform the ordinary least squares regression (Granger and Ramanathan, 1984):

$$y_{t+1} = a_0 + \sum_{j=1}^{k} a_k f_{t,j} + \varepsilon_{t+1} \tag{10.9}$$

where $f_{t,j}$ is the one step ahead prediction mad at time t of y_{t+1} with model i; a_0 is a constant term, and a_j is the regression coefficient.

10.4 Nonlinear Prediction Combination

Building a multi-model combination is a common way to improve the performance of the resulting model for classification and regression tasks. The combination model consists of a number of individual predictive models, such as: neural networks, support vector machines or regression trees. This introduces model diversity which is the central feature of the combination approach. The combination model often performs better than a single model based on the bias-variance decomposition of combination models (Krogh & Sollich, 1997). In our earlier work (Siek & Solomatine, 2010b) for combining models we used dynamic averaging and it is briefly described below. In this work, we propose to use a more sophisticated method, dynamic neural network, for the same purpose.

10.4.1 Dynamic averaging

Each individual predictive chaotic model predicts the future trajectory projection in phase space. Two linear prediction combination methods are utilized in this work: simple averaging and dynamic averaging. Simple averaging technique averages the predictions from all individual models with the same weights. On other hands, a dynamic averaging method incorporates the model selection and model combination procedures at once with some weights. The selection is measured by the dynamical performance (prediction accuracy) of each model. A penalty factor is used to remove low performing models (relatively compared to the best one) from the poll.

Let $e_{i,j}$ be the prediction error at time j generated by an combination model member i, the combination model member i is selected into the combination if the following both inequalities are satisfied:

$$\frac{1}{n}\sum_{i=1}^{n}\sum_{j=t-s+1}^{t}\left|e_{i,j}\right| - \min_{i=1}^{n}\left(\sum_{j=t-s+1}^{t}\left|e_{i,j}\right|\right) \leq pf \times \underset{i,j}{std}\left(e_{i,j}\right) \tag{10.10}$$

$$\frac{1}{s}\sum_{j=t-s+1}^{t}\left|e_j\right| \leq t_s \tag{10.11}$$

where n is the number of combination model members, t is the current time, s is the length of previous prediction errors to be moving averaged (i.e. $s=12$), std is the standard deviation, pf is a penalty factor (i.e. 2) and t_s is a threshold value (i.e. 10 cm). Hence, using this technique, the number of the selected models is different for each time step of prediction.

10.4.2 Dynamic neural networks

ANNs can be classified into dynamic and static categories. Static networks (e.g. MLP) have no feedback elements and contain no delays and the output is calculated directly from the input through feedforward connections. Its response at any given time depends not only on the current input, but on the history of the input sequence. Therefore, the dynamic network has memory (Haykin, 1999).

Dynamic networks are generally more powerful but more difficult to train than static networks. They can be trained to learn sequential or time-varying patterns. A more complex gradient-based algorithm for static networks can be used for training a dynamical network. The error surfaces for dynamic networks can be more complex than those for static networks and training is more likely to be trapped in local minima. There are two training types: batch and incremental training. In batch training, the weights and biases are updated after the entire training set has been applied to the network. Whereas in incremental training the weights and biases are updated each time an input is presented to the network. The batch training algorithm is generally much faster than the incremental (Fu *et al.*, 2002; Vo, 2002).

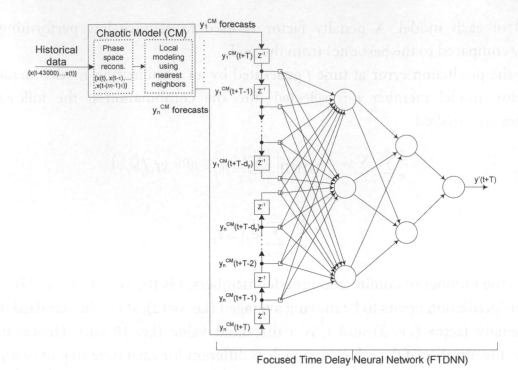

Figure 10-4: The architecture of a focused time delay neural network with tapped delay inputs fed from several predictive chaotic model predictions.

In this work, we use one type of dynamical networks, so-called Focused Time Delay Neural Network (FTDNN). The FTDNN consists of a feed-forward network with a tapped delay line at the input. Thus, the dynamics appear only at the input layer of a static multilayer feed-forward network. Figure 10-4 illustrates FTDNN architecture.

10.5 Model Results and Discussion

10.5.1 Global model

Back-propagation multi-layer perceptron (MLP) with Levenberg-Marquardt training rule (Haykin, 1999) was utilized and trained using the same input structure as the predictive chaotic model inputs (phase space reconstruction structure). The number of hidden neurons of ANN was selected using the exhaustive search in the range [1~10] and we found that four is the optimal number of hidden nodes.

10.5.2 Local model

Nonlinear analysis of surge time series recommends the appropriate values of time delay and embedding dimension are $\tau=10$ and $m=8$. We utilized sensitivity analysis to search for the appropriate number of neighbors (k) for non-stormy and stormy periods. The sensitivity analysis was conducted by setting up the predictive chaotic model parameters for the surges with $\tau=10$ and $m=8$ and the number of neighbors (k) run from 1 to 2000. We used 3rd-order polynomial local model as a reference.

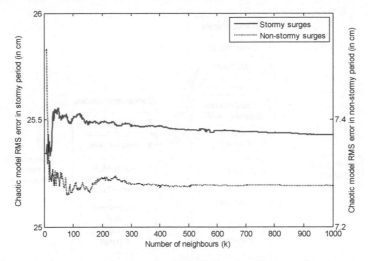

Figure 10-5: The six-hours ahead prediction error of the predictive chaotic models as a function of the number of neighbors (k) for surges during non-stormy and stormy periods ($\tau=10$, $m=8$).

Figure 10-5 depicts the six-hours prediction RMS errors of the predictive chaotic models as a function of the number of neighbors (k) for non-stormy and stormy surges. It is clearly shown that the suitable number of neighbors for predicting surges during storm condition is small (13 neighbors) and it should be smaller than the one (80 neighbors) during non-storm condition. One of the reasons is that less true dynamical neighbors (similar surge behavior in the past) can be found especially during extreme storms. If more neighbors are considered, the model performance will be worse due to the inclusion of false neighbors in constructing local models. Consequently, the whole predictive chaotic model performance will decrease. For MLP-NN local model, we set the number of neighbors to 300 for giving enough training dataset into the MLP-NN. The prediction horizons are 1, 3, 6, 10 and 12 hours. Each prediction horizon can have different values of time delay and embedding dimension. The result of optimization is the most accurate predictive chaotic model which has the lowest RMS error on cross validation data set. The cross validation data sets have small size of 400 data points: time indices of 35500-35900 for storm condition and 38200-38600 for non-storm condition.

The optimal time delay and embedding dimension for predictive chaotic model for predicting surges during storm condition was obtained (8 time-delayed input variables):

$$\mathbf{Y}_{t+T} = \{s_t^{hvh}, s_{t-10}^{hvh}, \dots, s_{t-70}^{hvh}\} \tag{10.12}$$

The performance of the local models is listed in Table 10-1. The RMS errors of 123-polynomials (linear, quadratic and cubic) are the same. This is due to the fact that the dynamical neighbors found for each prediction step are the same resulting in the same polynomial regressions.

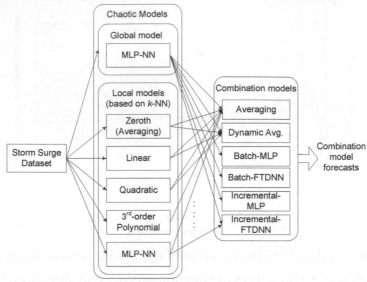

Figure 10-6: Schematic description of predictive chaotic models and prediction combination models in high dimensional chaotic system.

10.5.3 Dynamic averaging

A schematic description of the prediction combination models in high dimensional space is depicted in Figure 10-6. In the combination model, we have seven combination members: global MLP-NN, local models (zeroth, 123-Polynomials, MLP-NN) with direct and m-step predictions. For simple averaging technique, we average the predictions from all combination members with the same weights. In dynamic averaging technique, the moving average of the performances of the previous twelve predictions made by each combination members and a penalty factor (s=12; pf=2; t_s=10) are used for model selection and combination. These twelve predictions are chosen because the tides has a significant influence to the surge level variations and the dominant tidal constituent is M2 (principal lunar semidiurnal), which has about 12 hours tidal period/cycle.

10.5.4 Dynamic neural networks

The inputs of FTDNN are the time delay predictions from various types of predictive chaotic models, defined as:

$$\hat{y}_t^{comb} = f(y_t^{gMLP},...,y_{t-11}^{gMLP}, y_t^{zeroth},...,y_{t-11}^{zeroth}, y_t^{poly},...,$$
$$y_{t-11}^{poly}, y_t^{poly},...,y_t^{lMLP},...,y_{t-11}^{lMLP},....) \tag{10.13}$$

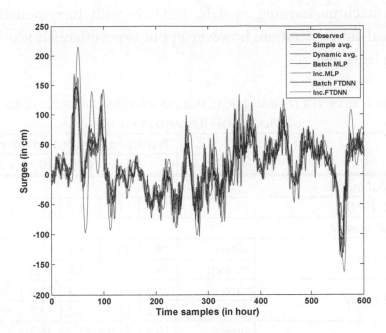

Figure 10-7: Performance comparison between several prediction combination techniques during stormy periods at Hoek van Holland station.

The prediction accuracy of multi-model ensembles is much higher compared to the ones of the global and local individual models (see Table 10-1 and Figure 10-7). The best performance is generally achieved by the dynamic neural network FTDNN with batch learning (in comparison to dynamic averaging and MLP-NN). This is due to the fact that this combination method allows for taking the best predictions at each step from a number of individual models which are selected by measuring the performance dynamics of each individual model. However, FTDNN does not perform well for short-term prediction (1 hour ahead). This might be due to fact that each individual model has already performed very well and these predictions are close to each other so that FTDNN has difficulty in identifying the behavior of prediction dynamics from each model member. Nevertheless, this combination technique performs well for long-term predictions. In spite of relatively lower performance, the incremental learning is advantageous for real-time operation and one of the ideas to explore here would be the "second-level" dynamic combination of incremental and batch learning.

10.6 Summary

Several combination techniques for a high dimensional chaotic system has been introduced and tested on a real-life case study. It can be concluded that combining different algorithms used in predictive chaotic models leads to improvements in accuracy. In the considered application to storm surge modeling, the Focused Time Delay Neural Network (FTDNN) has demonstrated the best capability for accurate prediction outperforming specialized (local) and global machine learning models. FTDNN with incremental learning can be recommended in real-time operation; however in our experiments it was less accurate than FTDNN with batch learning.

TABLE 10-1: PERFORMANCES OF THE GLOBAL, LOCAL MODELS AND ENSEMBLES IN STORM SURGE PREDICTIONS AT HOEK VAN HOLLAND STATION.

		Prediction horizons (in hours)				
		1	3	6	10	12
Global model						
MLP-NN		6.8	11.9	14.7	27.6	23.1
Local models						
Zeroth	direct	19.5	25.9	28.8	35.4	35.9
	m-step	19.5	31.2	35.9	37.8	38.0
123-Poly[*]	direct	11.6	21.8	25.4	35.0	36.0
	m-step	11.6	11.9	23.2	33.7	38.0
MLP-NN	direct	15.6	21.0	27.2	36.2	37.8
	m-step	15.6	21.0	27.2	36.2	37.8
Combination models						
Simple averaging		11.8	15.3	19.1	27.4	28.9
Dynamic avg.		6.7	11.0	16.3	30.0	28.7
Batch-MLP		**5.2**	9.0	10.1	23.2	19.4
Incremental-MLP		21.3	30.1	30.9	28.2	25.0
Batch-FTDNN		10.3	**2.6**	**2.7**	**2.7**	**2.9**
Incremental-FTDNN		47.3	27.0	26.9	24.7	26.9

[*]123-poly: linear, quadratic, cubic polynomial approximation functions

CHAPTER 11: CONCLUSIONS AND RECOMMENDATIONS

> *"We have the duty of formulating, of summarizing, and of communicating
> our conclusions, in intelligible form, in recognition of the right of other free
> minds to utilize them in making their own decisions."*
>
> Ronald Fisher

11.1 Main Conclusions

The methods of nonlinear dynamics, chaos theory and machine learning have been
established as a set of tools for modeling complex systems. Many of complex natural
phenomena have been studied and investigated by means of these techniques for better
understanding on how the nature works and to predict the future condition of nature. A
number of research shows that most of these natural phenomena can exhibit deterministic
chaos. The initial study on building a univariate chaotic model for predicting storm surges
in the North Sea has been conducted in 1999-2000 (reported by Solomatine *et. al.* (2000).
This chaotic model has been extended into a multivariate model (PhD study by Velickov
(2004), which can include other variables, such as wind and air pressure. The nonlinear
analysis of the observed time series indicates at the storm surge dynamics along the Dutch
coast can be characterized as deterministic chaos. Chaotic behavior in the storm surge
dynamics can be due to the fact that this system is the result of complex interactions
between various forces or dynamical systems, such as atmospheric dynamics, wind-wave-
tide interactions. The presence of deterministic chaos and positive Lyapunov exponent
implies the possibility for predictions. However, predictability of any model including
predictive chaotic model has some limits. Properties of the sensitivity to initial condition
and the existence of bifurcations can be the reasons of exponentially decreasing prediction
accuracy of chaotic model or any model, especially for long-term prediction. Yet, short and
medium term predictions are generally reliable and accurate.

In building a predictive chaotic model, the observed time series is reconstructed and embedded in sufficiently higher dimensional phase space with the time-delayed coordinates to unfold the attractor of surge dynamics. The predictive models can be constructed based on available data-driven techniques (i.e. ANN). Global and local models can be built. In global modeling, the whole dynamical behavior of the systems in phase space is described and predicted by one model. In contrast, the local modeling at each time step allows for characterizing the dynamical behavior locally and more flexible options of predictive local models can be utilized. However, this flexibility introduces a concern on selecting the good searching techniques for finding true dynamical neighbors and choosing the suitable number of dynamical neighbors used for building predictive local models.

If compared to the earlier studies by Solomatine *et al.* (2000) and Velickov (2003; 2004) (see in Chapter 1), this research has introduced several improved techniques and innovations, including using recent techniques for nonlinear time series analysis (i.e. Cao's method), considerable improvements in the algorithms for building predictive chaotic models (schemes to avoid false neighbors: utilizing multi-step prediction and trajectory based method, using ANN as a local model), possibility to build the chaotic model in case of incompleteness in the time series, reducing phase space dimension (an important issue for multivariate chaos), solving phase shift prediction error, optimizing predictive chaotic model using GA and ACCO, incorporating data assimilation using NARX neural network, and combining predictions from different types of chaotic models using dynamic averaging and dynamic neural network. Furthermore, the additional new data set is used in this work and the prediction performances of the predictive chaotic models are compared with other models, including ANN models.

A number of enhancements in building a predictive chaotic model outlined in the objectives of this research have been implemented and tested. The main conclusions can be summarized as follows:

- Taking into account the presence of deterministic chaos in surge dynamics, a mixture of multivariate predictive local models in the reconstructed phase-space of the dynamical system, which uses information from the real dynamical neighbors, has demonstrated a good capability for reliable short-term predictions. For the Hoek van Holland location, the overall 3 hours ahead surge prediction errors (RMSE) during storm condition for univariate CM, univariate ANN, multvariate CM and multivariate ANN are 12.91, 19.46, 11.99 and 16.78 cm, respectively.

- High dimensionality of multi-variable phase space prompts for employing dimensionality reduction methods. Method of phase space reconstruction of a dynamical system incorporating dimensionality reduction using principal component analysis (PCA) is presented. In the reduced dimensional phase space, multivariate predictive local models are built from the real dynamical neighbors. Surge data along the Dutch coast which can be characterized as deterministic chaos were obtained for testing our models. The results show that the use of dimensionality reduction method in the phase space reconstruction can improve the performance of univariate and multivariate predictive chaotic models outperforming ANN models. For the Hoek van Holland location, the overall prediction error for surges 10 hours ahead is about 5 cm and 14.5 cm for non-storm and storm conditions, respectively.

- Building a predictive chaotic model from incomplete time series is proposed in the view of possible failures of measurement instrument or data transmission in the real operation. Several imputing algorithms, such weighted sum of linear interpolation, Bayesian PCA and cubic spline interpolation are utilized. The resulting models are compared with the ones built from complete time series. The results indicate that the imputing techniques used can be incorporated into the predictive chaotic model for handling the missing values and the resulting models with imputation have still comparable performance with the models without missing values. The imputing technique of cubic spline interpolation generally outperforms the other techniques.

- An approach of correcting phase error in the chaotic model predictions is presented. Building a separate model for characterizing the phase error dynamics, such as predictive chaotic model and ANN is proposed. In the application of storm surge prediction, the predictive chaotic model and ANN model can identify and predict the dynamical behavior of the phase error generated by a standard chaotic model. In addition, they are able to automatically estimate and correct these phase prediction errors. This demonstrates that the proposed techniques can be used to enhance the predictability of a predictive chaotic model, e.g. for longer-term predictions.

- Identification and selection of proper dynamical neighbors in the reconstructed phase space are the major aspects in the local modeling approach. The presence of different regimes or similar behaviors in the dynamics can be an indicator for selecting different type of local models and number of neighbors used. The performance improvement of a predictive chaotic model is expected if the good searching algorithm is used for finding true neighbors. Besides the standard Euclidean distance method, a new method – the so-called trajectory based method – is proposed. The trajectory based method arises from an idea that finding true neighbors does not only depend on the distance between two points in the m-dimensional phase space, but also the distance and direction of two different trajectories (sequences of points in phase space) partly formed by these two points. The modeling results indicate that the trajectory based method is able to find true neighbors and the performance of predictive chaotic model using this technique is improved.

- Randomized search and and grid search appears to be viable methods for optimizing predictive chaotic model. Searching for the optimal values of embedding dimensions and time delay are important for unfolding the attractor of the surge dynamics and obtaining the smooth trajectories in phase space. Subsequently, it is crucial for building a reliable predictive chaotic model. Two optimization algorithm originated in the field of computational intelligence, the genetic algorithm (GA) and adaptive cluster covering (ACCO), are utilized for this purpose. The results shows that the application of optimization methods, grid search and randomized search can provide the better selection of the chaotic parameters in order to increase the performance of predictive chaotic models for storm surges. For the two case studies, the optimal sets of parameters found by using randomized search are: $\tau=16$, $m=28$, $k_{max}=22$ and RMSE=12.98 cm for the Hoek van Holland tidal station and $\tau=8$, $m=12$, $k_{max}=65$ and RMSE=3 cm for the San Juan tidal station (this data set was additionally used in the optimization experiments). The optimized models outperform the original model with parameters defined by nonlinear time series analysis in terms of prediction accuracy. The smoothness of the trajectories and unfolded attractor in phase space are improved as well, and this is indicated by the model performance.

- A real-time data assimilation technique using NARX neural network is introduced and implemented to re-analyze and improve predictive chaotic model predictions for storm surges in the North Sea. The predictive chaotic model with data assimilation has demonstrated a pronounced capability for reliable and accurate prediction outperforming standard predictive chaotic model, ANN model and the European numerical storm surge models. Our experiments show that for 48-hours ahead prediction at Hoek van Holland station during stormy period, the predictive chaotic models with NARX data assimilation with the frequency of every 6 hours (5.6cm RMS error) outperforms the standard predictive chaotic model (36.35cm RMS error) and the KNMI numerical model with EnKf data assimilation (11.62cm RMS error).

- Multi-model ensemble prediction using dynamic averaging and dynamic neural network model is introduced. The dynamic averaging technique allows for combining the individual model predictions on the basis of model performances in certain periods of prediction time. One type of dynamic neural network, so-called focused time-delayed neural network (FTDNN) is used. Several predictions from different types of predictive chaotic models are selected and further combined by these two techniques in order to obtain more accurate and reliable predictions. For a high-dimensional chaotic system, it means building an ensemble of all future trajectories in phase space, estimated by the heterogeneous individual models. In application to storm surge prediction, the FTDNN model has demonstrated higher capability for accurate prediction outperforming dynamic averaging and individual models (local and global models). The FTDNN with incremental learning is more recommended in real-time operation; however in our experiments it was less accurate than that with batch learning.

These modeling techniques based on the methods of nonlinear dynamics and chaos theory with several enhancements and innovations have demonstrated improved performance of predictive chaotic model and it can serve as an efficient tool for accurate and reliable short-term predictions (especially for predicting storm surges) in order to support decision-makers for flood prediction and ship navigation. Table 11-1 summarizes comparative performances of various models and their combinations (please note however that it is not always possible to make direct comparisons between different methods due to different model setups, ways of performing assimilation and data sets used).

11.2 Limitations and Recommendations

Several limitations and recommendations on using chaotic models for prediction with several proposed enhancements were identified and summarized, as follows:

- The first recommendation is to further explore the possibilities of combining the new approaches and enhancements presented in this work, for example, to use data assimilation for multi-variate models, with the ensemble models, etc. We have the possibility of testing only some of such combinations. The suggested enhancements cannot be course randomly combined: each of them has different purpose, theoretical background, advantages and disadvantages in particular conditions.

- The use of a chaotic model for prediction is generally reliable and accurate only for the dynamical systems which behaviors can be characterized as deterministic chaos. Nonlinear time series analysis needs to be used before building a predictive chaotic model. If there is an indication of chaos (which can be stronger or weaker), then it is sensible to build and apply the local prediction models in phase space described in this work. It should be noted that for many dynamical systems (like tides) which is characterized by strongly periodic sinusoidal components there may be no need to apply sophisticated chaos theory analysis as they can be very well predicted by linear models based on fast Fourier transform.

TABLE 11-1: PERFORMANCE IMPROVEMENTS BY USING SEVERAL IMPROVED METHODS.

Improved Methods	Models	RMSE (cm)	%	Descriptions
Building PCM	U-ANN*	19.46	0%	Stormy period 3 hours prediction ($\tau =1$, $m=7$, $k=13$)
	M-ANN	16.78	16%	
	U-PCM	11.94	63%	
	M-PCM	11.99	62%	
Phase space dimensionality reduction	U-ANN	13.09	53%	Stormy period 6 hours prediction ($\tau =$var, $m=$var, $k=9$-100)
	M-ANN*	20.05	0%	
	U-PCM	9.21	118%	
	M-PCM	12.16	65%	
	U-PCM with PCA	9.18	118%	
	M-PCM with PCA	13.69	46%	
Prediction error correction	U-PCM*	24.69	0%	Stormy period 3 hours prediction ($\tau =10$, $m=8$, $k=13$)
	U-PCM with PCM error corr.	12.74	94%	
	U-PCM with ANN error corr.	16.06	54%	
Incompleteness	U-PCM*	11	0%	Stormy period, 3 hours pred. 30% missing values ($\tau =10$, $m=8$, $k=13$) ($\tau =1$, $m=12$, $k=13$) for 2nd model
	U-PCM with weighted sum linear interp.	15.3	-28%	
	U-PCM with Bayesian PCA	12.9	-15%	
	U-PCM with cubic spline interp.	12	-8%	
Trajectory based method	U-PCM without cluster, Euclidean (Y3)*	3.7	0%	Stormy period, 6 hours pred. ($\tau =19$, $m=20$, $k=50$-100, $k=$var for Y6)
	U-PCM with cluster, Euclidean (Y1)	2.6	42%	
	U-PCM with trajectory based method (Y6)	1.3	185%	
Optimization	U-PCM ($\tau =3$, $m=6$, $k=13$), no cluster*	21.69	0%	Stormy period 6 hours prediction
	U-PCM with SES ($\tau =16$, $m=29$, $k=25$)	13.25	64%	
	U-PCM with GA ($\tau =24$, $m=20$, $k=26$)	13.75	58%	
	U-PCM with ACCO ($\tau =16$, $m=28$, $k=22$)	12.98	67%	
Data assimilation	U-PCM*	36.35	0%	Stormy period, 48 hours pred. New data (2003-2007) Data assimilation every 6 hours ($\tau =12$, $m=18$, $k=9$-100)
	U-ANN	33.74	8%	
	DCSM/WAQUA with EnKf	11.62	213%	
	U-PCM with NARX network	5.57	553%	
Multi-model ensembles	Global U-ANN	14.7	96%	Stormy period 6 hours prediction ($\tau =10$, $m=8$, $k=13$)
	Local U-ANN	27.2	6%	
	U-PCM Zeroth*	28.8	0%	
	U-PCM 123-Poly	23.2	24%	
	Ens-Simple Avg	19.1	51%	
	Ens-Dynamic Avg	16.3	77%	
	Ens-MLP with batch learning	10.1	185%	
	Ens-MLP with incremental learning	30.9	-7%	
	Ens-FTDNN with batch learning	2.7	967%	
	Ens-FTDNN with incremental learning	26.9	7%	

#PCM=Predictive chaotic model; U=univariate; M=multivariate; Ens=ensemble; *=reference model for calculating percentage

- Predictive chaotic model inevitably becomes less accurate for long-term prediction. However enhancements to increase the model predictability are possible, and several of them have been suggested, for example: using data assimilation scheme or combining different models.

- Finding the best chaotic model parameters (delay and embedding dimension) is computationally intensive task, and efficiency (speed) of the employed randomized search algorithms is of great importance. ACCO has been shown to be an efficient algorithm but it is recommended to investigate multi-algorithm schemes leading also to higher effectiveness (accuracy), combining for example, the employed GA and ACCO optimization algorithms with exhaustive search. A possibility is also to first using randomized search algorithm to find local optimal points and then refining the searching process using the exhaustive search around these local optimal points.

- A widely used data assimilation scheme, Ensemble Kalman Filter can be implemented for predictive chaotic model and compared its performance with data assimilation technique using NARX network.

- A major improvement in the accuracy of predictive chaotic model is expected if the meteorological predictions from numerical weather prediction model are directly used. For instance, the wind/pressure prediction fields (e.g. 48-hours predictions) can be utilized as additional inputs for predictive chaotic model. This technique can extend the predictability of chaotic model.

- More sophisticated methods to identify the adequate neighbors for building local models should be investigated further. It is also recommended to implement a mixture of various local models (like radial basis function), to employ smaller data sampling time or to construct non-equidistance phase space.

- The uncertainty analysis of chaotic model predictions (as of any predictive model) is an area for further research as well.

- The use of predictive chaotic model as a complementary model to the European operational storm surge models is highly recommended and would be the next step of this research.

- Other improvement can be achieved in the development of multi-model ensemble prediction techniques. The storm surge predictions from the European physically-based storm surge models for the North Sea from several meteorological institutions in the

Netherlands, Denmark, UK, Norway, Belgium and Germany and predictive chaotic model are combined by reliable multi-model ensemble techniques in order to obtain superior accuracy of predictions. In addition, the expert judgments can be included in the process of multi-model ensemble.

- The presented methodology ought to be tested on other cases studies, as more and more oceanographic data become available. We have already performed initial experiments for the San Juan case, and the results are encouraging.

- It is worth mentioning yet another observation about the limitation of using chaotic model for surge predictions – and it does not stem from the technical deficiencies of this approach. It relates to a long tradition of successful use of hydrodynamic models, the expertise and training of the specialists involved in such predictions, and their extremely busy schedules. It has been observed during this study, that in spite of the clear interest to the new techniques, making them really tested in operational environment and adopted requires much more than publications and presentations. It is quite natural and this is the case with many new technologies introduced in the public sector, especially in critical areas where human lives are at stake. It is therefore recommended not only to continue this research but also to continue undertaking efforts of making the advantages of the proposed methodology known to experts and decision makers.

References

Abarbanel, H. D. I. (1996). *Analysis of observed chaotic data*: Springer Verlag.

Abarbanel, H. D. I., Brown, R., Sidorowich, J. J., & Tsimring, L. S. (1993). The analysis of observed chaotic data in physical systems. *Reviews of Modern Physics, 65*(4), 1331-1392.

Abbott, M. B. (1991). *Hydroinformatics: information technology and the aquatic environment*: Avebury Technical.

Aha, D. W., Kibler, D., & Albert, M. K. (1991). Instance-based learning algorithms. *Machine Learning, 6*(1), 37-66.

Akaike, H. (1974). A new look at the statistical model identification. *IEEE Transactions on Automatic Control, 19*(6), 716-723.

Albano, A. M., Muench, J., Schwartz, C., Mees, A. I., & Rapp, P. E. (1988). Singular-value decomposition and the Grassberger-Procaccia algorithm. *Physical review A, 38*(6), 3017-3026.

Albano, A. M., Passamante, A., & Farrell, M. E. (1991). Using higher-order correlations to define an embedding window. *Physica D: Nonlinear Phenomena, 54*(1-2), 85-97.

Alexandersson, H., Schmith, T., Iden, K., & Tuomenvirta, H. (1998). Long-term variations of the storm climate over NW Europe. *The Global Atmosphere and Ocean System, 6*(2), 97-120.

Ashlock, D. (2006). *Evolutionary computation for modeling and optimization* (Vol. 200). New York Springer-Verlag Inc.

Babovic, V., Keijzer, M., & Bundzelm, M. (2000). *From global to local modelling: a case study in error correction of deterministic models*. In Proc. of the Fourth International Conference on Hydroinformatics, Rotterdam.

Barreto, G., & Araujo, A. (2001). *A self-organizing NARX network and its application to prediction of chaotic time series*. In Proc. of the IEEE Int. J. Conf. on Neural Networks, Washington, DC, USA.

Battjes, J. A., & Gerritsen, H. (2002). Coastal modelling for flood defence. *Phil. Trans. R. Soc. A, 360*(1796), 1461-1475.

Beniston, M., Stephenson, D., Christensen, O., Ferro, C., Frei, C., Goyette, S., et al. (2007). Future extreme events in European climate: an exploration of regional climate model projections. *Climatic Change, 81*(0), 71-95.

Bijl, W. (1997). Impact of a wind climate change on the surge in the southern North Sea. *Climate Research, 8*, 45-59.

Bishop, C. (2006). *Pattern recognition and machine learning*. New York: Springer Verlag.

Bode, L., & Hardy, T. A. (1997). Progress and recent developments in storm surge modeling. *Journal of Hydraulic Engineering, 123*(4), 315-331.

Bouttier, F., & Courtier, P. (1999). *Data assimilation concepts and methods* (Vol. 14): ECMWF Meteorological Training Course Lecture Series.

Box, G., Jenkins, G., & Reinsel, G. (1994). *Time series analysis: forecasting and control* (3rd ed.): Prentice Hall, NJ, USA.

Breiman, L. (1984). *Classification and regression trees*: Chapman & Hall/CRC.

Broersen, P. M. T. (2006). *Automatic autocorrelation and spectral analysis*: Springer-Verlag New York Inc.

Bryson, A., & Ho, Y. Applied Optimal Control. 1969. *Blaisdell, Waltham, Massachusetts*.

Bryson, A., & Ho, Y. (1969). *Applied Optimal Control*.

Burnham, K., & Anderson, D. (2002). *Model selection and multimodel inference: a practical information-theoretic approach*: Springer Verlag.

Butler, A., Heffernan, J. E., Tawn, J. A., Flather, R. A., & Horsburgh, K. J. (2007). Extreme value analysis of decadal variations in storm surge elevations. *Journal of Marine Systems, 67*(1-2), 189-200.

Buzug, T., & Pfister, G. (1992). Optimal delay time and embedding dimension for delay-time coordinates by analysis of the global static and local dynamical behavior of strange attractors. *Physical review A, 45*(10), 7073.

Cao, L. (1997). Practical method for determining the minimum embedding dimension of a scalar time series. *Physica D: Nonlinear Phenomena, 110*(1-2), 43-50.

Carter, R. W. G. (1985). North Sea dynamics. *Earth-Science Reviews, 22*(1), 103-104.

Casdagli, M. (1989). Nonlinear prediction of chaotic time series. *Physica D: Nonlinear Phenomena, 35*(3), 335-356.

Castillo, E., Hadi, A. S., Alegría, J. M. S., & Balakrishnan, N. (2005). *Extreme value and related models with applications in engineering and science*: Wiley Chichester.

Cleveland, W. S. (1979). Robust locally weighted regression and smoothing scatterplots. *Journal of the American statistical association*, 829-836.

Corkan, R., & Council, L. (1948). Storm surges in the North Sea. *U.S. Hydrogr. Off., Misc. 15072, Vol. 1, 174 pp. and Vol. 2, 166 pp.*

Crutchfield, J., Schuster, P., & Corporation, E. (2003). *Evolutionary dynamics: exploring the interplay of selection, accident, neutrality, and function*: Oxford University Press New York.

de Vries, H., Breton, M., de Mulder, T., Krestenitis, Y., Ozer, J., Proctor, R., et al. (1995). A comparison of 2D storm surge models applied to three shallow European seas. *Environmental Software, 10*(1), 23-42.

de Vries, J. (1991). *The implementation of the WAQUA/CSM16 model for real time storm surge forecasting*. de Bilt: KNMI.

Debernard, J., SÃ¦tra, Ã. y., & RÃ¸ed, L. P. (2002). Future wind, wave and storm surge climate in the northern North Atlantic. *Climate Research, 23*(1), 39-49.

Deltares. (2010). *Delft3d-FLOW: Simulation of multi-dimensional hydrodynamic flows and transport phenomena, including sediments*. Delft: Deltares.

Deo, M. (2010). Artificial neural networks in coastal and ocean engineering. *IJMS, 39*, 589-596.

Dercole, F., & Rinaldi, S. (2008). *Analysis of evolutionary processes: the adaptive dynamics approach and its applications*: Princeton Univ Pr.

Dijkstra, H. (2005). *Nonlinear physical oceanography: a dynamical systems approach to the large scale ocean circulation and El Niño*: Springer.

Donner, R., & Barbosa, S. (2008). *Nonlinear Time Series Analysis in the Geosciences*. Berlin: Springer.

Dronkers, J. J. (1964). *Tidal computations in rivers and coastal waters* (Vol. 530).

Droppert, L. (2001). *The NOOS Plan: North West Shelf Operational Oceanographic System 2002-2006*: Southampton Oceanography Centre.

Eckmann, J. P., Kamphorst, S. O., & Ruelle, D. (1987). Recurrence plots of dynamical systems. *EPL (Europhysics Letters), 4*, 973.

Eckmann, J. P., & Ruelle, D. (1985). Ergodic theory of chaos and strange attractors. *Reviews of Modern Physics, 57*(3), 617.

El Serafy, G., Gerritsen, H., Hummel, S., Weerts, A., Mynett, A., & Tanaka, M. (2007). Application of data assimilation in portable operational forecasting systems—The DATools assimilation environment. *Ocean Dynamics, 57*(4), 485-499.

Emery, W., & Thomson, R. (2001). *Data analysis methods in physical oceanography*: Elsevier Science Ltd.

Engelbrecht, A. P. (2007). *Computational intelligence: an introduction*: Wiley.

Evensen, G. (2007). *Data assimilation: The ensemble Kalman filter*: Springer Verlag.

Farmer, J. D., & Sidorowich, J. J. (1987). Predicting chaotic time series. *Physical Review Letters, 59*(8), 845-848.

Fermi, E., Pasta, J., & Ulam, S. (1955). *Studies of nonlinear problems*. Los Alamos: Los Alamos Scientific Laboratory.

Fix, E., & Hodges, J. (1951). *Discriminatory analysis*. Texas: USAF School of Aviation Medicine.

Fraser, A., & Swinney, H. (1986). Independent coordinates for strange attractors from mutual information. *Physical review A, 33*(2), 1134-1140.

Fraser, A. M. (1989). Information and entropy in strange attractors. *IEEE Transactions on Information Theory, 35*(2), 245-262.

Fraser, A. S. (1958). Monte Carlo analyses of genetic models. *Nature, 181*, 208 - 209.

Friedman, J. H. (1991). Multivariate adaptive regression splines. *The annals of statistics*, 1-67.

Frison, T., Abarbanel, H., Earle, M., Schultz, J., & Scherer, W. (1999). Chaos and predictability in ocean water levels. *J. Geophys. Res., 104*.

Fu, L. M., Hsu, H. H., & Principe, J. C. (2002). Incremental backpropagation learning networks. *IEEE Transactions on Neural Networks, 7*(3), 757-761.

Furtado, H. C. M., Velho, H. F. d. C., & Macau, E. E. N. (2008). Data assimilation: Particle filter and artificial neural networks. *Journal of Physics: Conference Series*, 012073.

Gaspard, P. (1998). *Chaos, scattering, and statistical mechanics* (Vol. 9): Cambridge Univ Pr.

Gerritsen, H., De Vries, H., & Philippart, M. (1995). The Dutch continental shelf model. *Coastal and Estuarine Studies*, 425-425.

Ghilardi, P., & Rosso, R. (1990). Comment on "Chaos in Rainfall" by I. Rodriguez-Iturbe et al. *Water Resources Research, 26*(8), 1837-1839.

Glieck, J. (1987). Chaos: Making a new science. *Viking, New York*.

Gonnert, G., Dube, S., Murty, T., & Siefert, W. (2001). Global storm surges: Theory observation and application. *German Coastal Engineering Research Council, 623*.

Grassberger, P., & Procaccia, I. (1983a). Characterization of strange attractors. *Physical Review Letters, 50*(5), 346-349.

Grassberger, P., & Procaccia, I. (1983b). Measuring the strangeness of strange attractors. *Physica D: Nonlinear Phenomena, 9*(1-2), 189-208.

Han, M., & Fan, M. (2006). *Multivariate Time Series Prediction by Neural Network Combining SVD.* In Proc. of the IEEE Int. Conf. on Systems, Man and Cybernetics, Taipei.

Härter, F. P., & de Campos Velho, H. F. (2008). New approach to applying neural network in nonlinear dynamic model. *Applied Mathematical Modelling, 32*(12), 2621-2633.

Hasan, S. (2001). *Exploratory analysis of tidal current in the Maas channel.* M.Sc. Thesis, Hydroinformatics, UNESCO-IHE, Delft.

Haykin, S. (1999). *Neural networks: a comprehensive foundation*: Prentice Hall.

Haykin, S., Principe, J., Sejnowski, T., McWhirter, J., & Cambridge, M. (2007). *New directions in statistical signal processing: from systems to brain*: MIT Press.

Heemink, A., Bolding, K., & Verlaan, M. (1997). *Storm surge forecasting using Kalman filtering.* In Proc.

Heemink, A. W., & Metzelaar, I. D. M. (1995). Data assimilation into a numerical shallow water flow model: a stochastic optimal control approach. *Journal of Marine Systems, 6*(1-2), 145-158.

Hegger, R., Kantz, H., & Schreiber, T. (1999). Practical implementation of nonlinear time series methods: The TISEAN package. *Chaos: An Interdisciplinary Journal of Nonlinear Science, 9*(2), 413-435.

Hirsch, M., Smale, S., & Devaney, R. (2004). *Differential equations, dynamical systems, and an introduction to chaos*: Academic Pr.

Hochman, M. (2011). Dynamical Systems Theory: What in the World is it? Retrieved from http://math.huji.ac.il/~mhochman/research-expo.html

Holland, J. H. (1975). Adaptation in natural and artificial systems.

Holthuijsen, L. (2007). *Waves in oceanic and coastal waters*: Cambridge Univ Pr.

Horne, B., & Giles, C. (1995). An experimental comparison of recurrent neural networks. *Advances in Neural Information Processing Systems*, 697-704.

Horsburgh, K. J., & Wilson, C. (2007). Tide-surge interaction and its role in the distribution of surge residuals in the North Sea. *J. Geophys. Res., 112.*

Howarth, M. J., John, H. S., Karl, K. T., & Steve, A. T. (2001). North Sea Circulation. *Encyclopedia of Ocean Sciences* (pp. 1912-1921). Oxford: Academic Press.

Hsieh, W. (2009). Machine Learning Methods in the Environmental Sciences. *Cambridge Univ. Pr., Cambridge.*

Ivars, P. (1993). Newton's Clock: Chaos in the Solar System: WH Freeman.

Jamieson, D., Wilkinson, J., & Ibbitt, R. (1972). *Hydrologic forecasting with sequential deterministic and stochastic stages.* In Proc. of the Int. Symp. on Uncertainties in Hydrologic and Water Res. Syst., Tucson, Arizona.

Jang, J., Sun, C., & Mizutani, E. (1997). *Neuro-Fuzzy and Soft Computing-A Computational Approach to Learning and Machine Intelligence.* NJ: Prentice Hall.

Jayawardena, A., & Lai, F. (1994). Analysis and prediction of chaos in rainfall and stream flow time series. *Journal of Hydrology, 153*(1-4), 23-52.

Jiayu, L., Yueke, W., Zhiping, H., & Zhenkang, S. (2006). Selection of Proper Time-Delay in Phase Space Reconstruction of Speech Signals. *Front. Electr. Electron. Eng. China, 1*, 111–114.

Jonker, H., & van Reeuwijk, M. (2010). *Chaotic processes.*Unpublished manuscript, Delft.

Kalnay, E. (2003). *Atmospheric modeling, data assimilation, and predictability*: Cambridge Univ Press.

Kantz, H., & Schreiber, T. (2004). *Nonlinear time series analysis*: Cambridge Univ Pr.

Kaplan, J. L., & Yorke, J. A. (1979). Chaotic behavior of multidimensional difference equations. *Lecture notes in mathematics, 730*, 204-227.

Kennel, M., Brown, R., & Abarbanel, H. (1992). Determining embedding dimension for phase-space reconstruction using a geometrical construction. *Physical review A, 45*(6), 3403-3411.

Kolmogorov, A. (1958). A new metric invariant of transitive dynamical systems and Lebesgue space automorphisms. *Dokl. Akad. Nauk. SSSR, 119*, 861-864.

Korteweg, D. J., & de Vries, G. (1895). On the change of form of long waves advancing in a rectangular canal and on a new type of long stationary waves. *Phil. Mag., 39*, 422-443.

Krogh, A., & Sollich, P. (1997). Statistical mechanics of ensemble learning. *Physical Review E, 55*(1), 811-825.

Kugiumtzis, D., Lingjaerde, O. C., & Christophersen, N. (1998). Regularized local linear prediction of chaotic time series. *Physica D: Nonlinear Phenomena, 112*(3-4), 344-360.

Langenberg, H., Pfizenmayer, A., von Storch, H., & Sündermann, J. (1999). Storm-related sea level variations along the North Sea coast: natural variability and anthropogenic change. *Continental Shelf Research, 19*(6), 821-842.

Lee, J., & Verleysen, M. (2007). *Nonlinear dimensionality reduction*: Springer Verlag.

Li, Y. (2004). *Chaos in partial differential equations*: International Press.

Li, Y. (2007). Chaos in Partial Differential Equations, Navier-Stokes Equations and Turbulence. *Arxiv preprint arXiv:0712.4026*.

Lin, T., Giles, C., Horne, B., & Kung, S. (1997). A delay damage model selection algorithm for NARX neural networks. *IEEE Transactions on Signal Processing, 45*(11), 2719-2730.

Lorenz, E. N. (1963). Deterministic nonperiodic flow. *Journal of the Atmospheric Sciences, 20*(2), 130-141.

Lundberg, A. (1982). Combination of a conceptual model and an autoregressive error model for improving short time forecasting. *Nordic Hydrology, 13*(4), 233-246.

MacKay, D. (2003). *Information theory, inference, and learning algorithms*: Cambridge Univ Pr.

Marwan, N., Romano, M., Thiel, M., & Kurths, J. (2007). Recurrence plots for the analysis of complex systems. *Physics Reports, 438*(5-6), 237-329.

McCulloch, W. S., & Pitts, W. (1943). A logical calculus of the ideas immanent in nervous activity. *Bulletin of Mathematical Biology, 5*(4), 115-133.

McInnes, K. L., Walsh, K. J. E., Hubbert, G. D., & Beer, T. (2003). Impact of Sea-level Rise and Storm Surges on a Coastal Community. *Natural Hazards, 30*(2), 187-207.

Menezes, J., & Barreto, G. (2008). Long-term time series prediction with the NARX network: An empirical evaluation. *Neurocomputing, 71*(16-18), 3335-3343.

Miles, J. W. (1957). On the generation of surface waves by shear flows. *Journal of Fluid Mechanics, 3*(02), 185-204.

Mitchell, T. (1997). *Machine learning*: McGraw Hill.

Mosteller, F. (1948). A k-Sample Slippage Test for an Extreme Population. *The Annals of Mathematical Statistics, 19*(1), 58-65.

Murcia, O. H. (2009). *Analysis and optimization of chaotic models for storm surge prediction*. M.Sc. Thesis, Hydroinformatics, UNESCO-IHE, Delft.

Nasell, I. (2001). Extinction and Quasi-stationarity in the Verhulst Logistic Model* 1. *Journal of Theoretical Biology, 211*(1), 11-27.

Ott, E., Sauer, T., & Yorke, J. (1994). *Coping with Chaos Analysis of Chaotic Data and the Exploitation of Chaotic Systems*. New York: John Wiley.

Otto, L., Zimmerman, J. T. F., Furnes, G. K., Mork, M., Saetre, R., & Becker, G. (1990). Review of the physical oceanography of the North Sea. *Netherlands Journal of Sea Research, 26*(2-4), 161-238.

Packard, N. H., Crutchfield, J. P., Farmer, J. D., & Shaw, R. S. (1980). Geometry from a time series. *Physical Review Letters, 45*(9), 712-716.

Palmer, T., Doblas-Reyes, F., Hagedorn, R., & Weisheimer, A. (2005). Probabilistic prediction of climate using multi-model ensembles: from basics to applications. *Philosophical Transactions of the Royal Society B: Biological Sciences, 360*(1463), 1991.

Peeck, H. H., Proctor, R., & Brockmann, C. (1982). Operational storm surge models for the North Sea. *Continental Shelf Research, 2*(4), 317-329.

Pikovsky, A., Author, Rosenblum, M., Kurths, J., Hilborn, R. C., & Reviewer. (2002). Synchronization: A Universal Concept in Nonlinear Science. *American Journal of Physics, 70*(6), 655-655.

Poincaré, H. (1890). Sur le problème des trois corps et les équations de la dynamique. *Acta Mathematica, 13*(1), 3-270.

Poincaré, H. (1952). *Science and Method: Translated by Francis Maitland*: Dover Publications.

Poincaré, H., & Halsted, G. B. (1913). *The Foundations of Science: science and hypothesis, the value of science, science and method*: The Science Press.

Prandle, D., & Wolf, J. (1978). The interaction of surge and tide in the North Sea and River Thames. *Geophysical Journal of the Royal Astronomical Society, 55*(1), 203-216.

Prandle, D., Wolf, J., & Jacques, C. (1978). Surge-tide interaction in the Southern North Sea. *Elsevier Oceanography Series* (Vol. 23, pp. 161-185): Elsevier.

Price, R. (2001). Hydroinformatics, modelling and knowledge management. *White paper, IHE-Delft*.

Proctor, R. (1995). *North Sea Model Advection-Dispersion Study NOMADS*. UK: Proj. Rep. Second MAST Days Euromar Market.

Provanzale, A., Smith, L., Vio, R., & Murante, G. (1992). Distinguishing between low-dimensional dynamics and randomness in measured time series. *Physica D, 58*(1-4), 31-49.

Pupo, C. J. R. (2000). *Application of data mining techniques and chaos theory for surge water level predictions (Hook of Holland case study)*. M.Sc. Thesis, Hydroinformatics, UNESCO-IHE, Delft.

Quinlan, J. R. (1992). *Learning with continuous classes*. In Proc. of the 5th Australian joint conference on artificial intelligence, Singapore.

Raftery, A. E., Gneiting, T., Balabdaoui, F., & Polakowski, M. (2005). Using Bayesian model averaging to calibrate forecast ensembles. *Monthly Weather Review, 133*(5), 1155-1174.

Rhode, C. (2011). Intro Neural Networks Retrieved from http://lowercolumbia.edu/students/academics/facultyPages/rhode-cary/intro-neural-net.htm

Robaczewska, K. B., Heemink, A. W., Verlaan, M., J.H. Stel, H. W. A. B. J. C. B. L. J. D., & Meulen, J. P. v. d. (1997). Data assimilation in the continental shelf model. *Elsevier Oceanography Series* (Vol. 62, pp. 472-482): Elsevier.

Rodriguez-Iturbe, I., De Power, B. F., Sharifi, M. B., & Georgakakos, K. P. (1989). Chaos in rainfall. *Water Resources Research, 25*(7), 1667-1675.

Roelvink, D., Reniers, A., van Dongeren, A., van Thiel de Vries, J., McCall, R., & Lescinski, J. (2009). Modelling storm impacts on beaches, dunes and barrier islands. *Coastal Engineering, 56*(11-12), 1133-1152.

Rosenstein, M., Collins, J., & Luca, C. (1994). Reconstruction expansion as a geometry-based framework for choosing proper delay times. *Physica-Section D, 73*(1), 82-98.

Ruelle, D. (1990). Deterministic chaos: the science and the fiction. *Proc. R. Soc. Lond. A, 427*, 241-248.

Rumelhart, D. E., Hintont, G. E., & Williams, R. J. (1986). Learning representations by back-propagating errors. *Nature, 323*(6088), 533-536.

Sano, M., & Sawada, Y. (1985). Measurement of the Lyapunov spectrum from a chaotic time series. *Physical Review Letters, 55*(10), 1082-1085.

Sauer, T., Yorke, J. A., & Casdagli, M. (1991). Embedology. *Journal of Statistical Physics, 65*(3), 579-616.

Schertzer, D., Tchiguirinskaia, I., Lovejoy, S., Hubert, P., Bendjoudi, H., & Larcheveque, M. (2002). Which chaos in the rainfall–runoff process? *Hydrological Sciences-Journal-des Sciences Hydrologiques, 47*(1).

Scott, A. (2005). *Encyclopedia of nonlinear science*: Routledge.

Shrestha, D., & Solomatine, D. (2006). Experiments with AdaBoost. RT, an improved boosting scheme for regression. *Neural computation, 18*(7), 1678-1710.

Siegelmann, H., Horne, B., & Giles, C. (1997). Computational capabilities of recurrent NARX neural networks. *IEEE Trans. on Systems, Man, and Cybernetics, Part B, 27*(2), 208-215.

Siek, M., & Solomatine, D. (2010a). Nonlinear chaotic model for predicting storm surges. *Nonlinear Processes in Geophysics, 17*, 405-420.

Siek, M., & Solomatine, D. P. (2010b). *Multi-model ensemble forecasting in high dimensional chaotic system*. In Proc. of the Int. Joint Conf. on Neural Networks, Barcelona.

Simonnet, E., Dijkstra, H. A., & Ghil, M. (2009). Bifurcation analysis of ocean, atmosphere and climate models. In R. M. Temam & J. J. Tribbia (Eds.), *Computational Methods for the Ocean and the Atmosphere*.

Sinai, Y. G. (1959). On the concept of entropy of a dynamical system. *Dokl. Akad. Nauk. SSSR, 124*, 768-771.

Sivakumar, B. (2004). Chaos theory in geophysics: past, present and future. *Chaos, Solitons & Fractals, 19*(2), 441-462.

Sivakumar, B., Liong, S. Y., Liaw, C. Y., & Phoon, K. K. (1999). Singapore rainfall behavior: chaotic? *Journal of Hydrologic Engineering, 4*, 38.

Smith, L. (1988). Intrinsic limits on dimension calculations. *Physics Letters A, 133*, 283-288.

Solomatine, D. (1999). Two strategies of adaptive cluster covering with descent and their comparison to other algorithms. *Journal of Global Optimization, 14*(1), 55-78.

Solomatine, D., Rojas, C., Velickov, S., & Wust, H. (2000). *Chaos theory in predicting surge water levels in the North Sea*. In Proc. of the Int. Conf. on Hydroinformatics, Iowa, USA.

Solomatine, D. P. (2002). *Data-driven modelling: paradigm, methods, experiences*. In Proc.

Solomatine, D. P., & Siek, M. B. (2006). Modular learning models in forecasting natural phenomena. *Neural networks, 19*(2), 215-224.

Solomatine, D. P., Velickov, S., & Wüst, J. C. (2001). *Predicting water levels and currents in the North Sea using chaos theory and neural networks*. In Proc. of the XXXIX IAHR Congress, Beijing.

Stewart, R. (2002). Introduction to physical oceanography. *Texas A&M University*.

Strogatz, S. (2001). *Nonlinear dynamics and chaos: With applications to physics, biology, chemistry, and engineering*: Perseus Books.

Szöllösy-Nagy, A., Bartha, P., & Harkányi, K. (1983). *Microcomputer Based Operational Hydrological Forecasting System for River Danube*. In Proc. of the Technical Conference on Mitigation of Natural Hazards through Real-Time Data Collection Systems and Hydrological Forecasting, Sacramento, CA.

Takens, F. (1981). Detecting strange attractors in turbulence. *Dynamical Systems and Turbulence* (Vol. 898, pp. 366-381). Berlin: Springer.

Takeuchi, K. (1976). Distribution of informational statistics and a criterion of model fitting. *Suri-Kagaku (Mathematical Sciences), 153*, 12-18.

Theiler, J. (1990). Estimating fractal dimension. *J. Opt. Soc. Am. A, 7*(6), 1055-1073.

Tsonis, A., & Elsner, J. (1988). The weather attractor over very short timescales. *Nature, 333*, 545-547.

Tsonis, A. A. (1992). *Chaos: from theory to applications*. New York: Plenum Press

Turing, A. M. (1950). Computing machinery and intelligence. *Mind, 59*(236), 433-460.

Unden, P., Rontu, L., Järvinen, H., Lynch, P., Calvo, J., Cats, G., et al. (2002). *HIRLAM-5 scientific documentation*. Sweden: HiRLAM 5 Project.

Velickov, S. (2004). *Nonlinear dynamics and chaos with applications to hydrodynamics and hydrological modelling*: Taylor & Francis Group.

Velickov, S., Price, R. K., & Solomatine, D. P. (2003). *Prediction of nonlinear dynamical systems based on time series analysis: Issues of entropy, complexity and predictability*. In Proc. of the XXX IAHR Congress, Thessaloniki, Greece.

Verhulst, P. (1845). Recherches mathématiques sur la loi d'accroissement de la population. Nouv. Mém. Acad. R. Sci. *Belles-Lettres Bruxelles, 18*, 1-41.

Verlaan, M., & Heemink, A. (1997). Tidal flow forecasting using reduced rank square root filters. *Stochastic Hydrology and Hydraulics, 11*(5), 349-368.

Verlaan, M., Zijderveld, A., de Vries, H., & Kroos, J. (2005). Operational storm surge forecasting in the Netherlands: developments in the last decade. *Phil. Trans. R. Soc. A, 363*(1831), 1441-1453.

Vitrano, J. B., & Povinelli, R. J. (2001). *Selecting dimensions and delay values for a time-delay embedding using a genetic algorithm.* In Proc. of the Genetic Evolutionary Computation Conference, San Francisco, USA.

Vo, M. T. (2002). *Incremental learning using the time delay neural network.* In Proc. of the Acoustics, Speech, and Signal Processing, Adelaide.

Walton, T. L. (2005). Short term storm surge forecasting. *Journal of Coastal Research, 21*(3), 421-429.

Wang, Y., & Witten, I. (1997). *Inducing model trees for continuous classes.* In Proc. of the Ninth European Conference on Machine Learning, Prague, Czech Republic.

Werbos, P. (1974). *Beyond regression: new tools for prediction and analysis in the behavioral sciences.* Unpublished PhD thesis, Harvard University.

Whitney, H. (1936). Differentiable manifolds. *The Annals of Mathematics, 37*(3), 645-680.

Wichard, J., & Ogorzalek, M. (2004). *Time series prediction with ensemble models.* In Proc. of the Int. J. Conf. on Neural Networks, Budapest.

Witten, I., & Frank, E. (2002). Data mining: practical machine learning tools and techniques with Java implementations. *ACM SIGMOD Record, 31*(1), 76-77.

WMO. (1992). *Simulated real-time intercomparison of hydrological models*: World Meteorological Organization.

WMO, S. (1998). Guide to Wave Analysis and Forecasting: WMO Publications.

WXTide32. (2009). WXTide32 - a free Windows tide and current prediction program. Retrieved from http://www.wxtide32.com/

Zabusky, N. J., & Kruskal, M. D. (1965). Interaction of" solitons" in a collisionless plasma and the recurrence of initial states. *Physical Review Letters, 15*(6), 240-243.

Zamani, A., Solomatine, D., Azimian, A., & Heemink, A. (2008). Learning from data for wind-wave forecasting. *Ocean Engineering, 35*(10), 953-962.

Zhu, Y. (2005). Ensemble forecast: A new approach to uncertainty and predictability. *Advances in atmospheric sciences, 22*(6), 781-788.

Michael Siek was born in Jember, East Java, Indonesia on August 14th, 1974. He received and remembers the first information from surroundings during his 1st birthday and in the age of five he started to realize of being as a small creature living in the chaotic world. In the following years, he became very curious and actively asking many questions to his parents and teachers, which might be foolish ideas and unanswered questions. One answer that he always remembers is the one from his father about Einstein's relativity theory. His father's explanation was unbelievably straightforward even in that time he was still in elementary school. After his father passed away, his interests on understanding nature and life did not stop but gradually emerged. At early time of his high school education, he and his elder brother won a national research competition and were invited to meet President and Minister of Education. Their research was about building a communication device using modulation effect of laser rays as transmission medium, which in that time this technology had not been invented yet.

On continuing his study to university, he took two different majors in two different universities at the same time. He earned his A.Md. degree in Information Management and Computer Engineering (with magna cumlaude) from STTS Institute in 1996, B.Sc. degree in Mathematics from Airlangga University (first upper class) in 2000, B.Com. degree in Information Management (with summa cumlaude) from Stikom Institute in 2000, all located in Surabaya, Indonesia. He appeared in most of newspapers and magazines in the nation for his excellent final thesis projects and being the best graduate. During his studies, he also worked at the universities as an assistant lecturer and laboratory staff as well as at several commercial companies as software developer, system analyst and consultant. After graduated from his university studies, he held positions in academia as a full-time lecturer in the Faculties of Engineering and Economics at University of Surabaya and a visiting lecturer at Petra Christian University. Few years later, he attained a Netherlands Fellowships Programme (NFP) scholarship and came to the Netherlands taking a M.Sc. programme in Hydroinformatics at UNESCO-IHE Institute for Water Education. Under supervision of Professor Dimitri Solomatine, he did M.Sc. research on "Optimality and flexibility of model trees in applications to water-related problems" and received his M.Sc. degree in 2003. This line of research was continued for another year and several papers have been published. After that he came back to Indonesia and took a position as Head of Computer Systems Department at a large company with responsibilities on executing several projects aimed at developing an integrated information system and implementing ERP software. During this time, he also did a joint research project between TU Eindhoven and Océ /Phillips on Bayesian network model for machine maintenance prediction.

In 2006, he went back to the Netherlands and started doing his Ph.D. research entitled "Predicting storm surges: chaos, computational intelligence, data assimilation, ensembles" with supervision of Professor Dimitri Solomatine at UNESCO-IHE and TU Delft. His research is partly supported by Delft Cluster Project on "Safety against flooding". He co-authored several papers published in the international journals and conference proceedings. He obtained best student presentation awards from the IEEE Computational Intelligence Society (IEEE-CIS) and the International Neural Network Society (INNS), and the European Geosciences Union (EGU) young scientist travel award. Besides doing research, he was also giving lectures and co-supervised M.Sc. student research at UNESCO-IHE.

His research interests have spanned a large number of disciplines, emphasizing the data-driven and physically-based modeling, nonlinear dynamics and chaos theory, computational intelligence, flood modeling, GIS, neural networks, data mining, optimization techniques, data assimilation, multi-model ensemble prediction with a wide range of real-life applications, such as the hydrological and coastal applications.

He is currently a member of the International Association of Hydrological Sciences (IAHS), International Association of Hydraulic Engineering and Research (IAHR), IEEE Computational Intelligence Society, International Neural Network Society (INNS), European Geosciences Union (EGU), Institute Mathematical Statistics (IMS) and European Network for Business and Industrial Statistics (ENBIS).

www.siek.biz

In international peer-reviewed journals

Siek, M.B. and Solomatine, D.P. (2011). Real-time Data Assimilation for Chaotic Storm Surge Model Using NARX Neural Network. *Journal of Coastal Research*, SI 64, 1184-1188.

Siek, M.B. and Solomatine, D.P. (2011). Optimized Dynamic Ensembles of Multiple Chaotic Models in Predicting Storm Surges. *Journal of Coastal Research*, SI 64, 1189-1194.

Siek, M. and Solomatine, D.P. (2010). Nonlinear Chaotic Model for Predicting Storm Surges. *Nonlinear Processes in Geophysics*, 17, 405-420.

Solomatine, D.P. and Siek, M.B. Modular learning models in prediction natural phenomena. *Neural Networks J.*, 2006, 19(2), 215-224.

In proceedings of the international conferences

Siek, M.B. and Solomatine, D.P. (2011). *Nonlinear Multi-model Ensemble Prediction Using Dynamic Neural Network with Incremental Learning.* In Proceedings of the IEEE International Joint Conferences on Neural Networks, San Jose, USA (Best Student Presentation Award).

Siek, M.B. and Solomatine, D.P. (2011). *Predicting Ocean Surge: Optimized Ensembles of Data Driven Chaos-based Models in Phase Space.* In Proceedings of the 34th IAHR World Congress, Brisbane, Australia.

Siek, M.B. and Solomatine, D.P. (2010). *Phase Error Correction for Chaotic Storm Surge Model.* In Proceedings of the International Conference on Hydroinformatics, Tianjin, China.

Siek, M.B. and Solomatine, D.P. (2010). *A Mixture of Multi-Neural Networks in Phase Space Reconstruction.* In Proceedings of the International Conference on Hydroinformatics, Tianjin, China.

Siek, M.B. and Solomatine, D. P. (2010). *Multi-models Ensemble Prediction in High Dimensional Chaotic System.* In Proceedings of the IEEE International Joint Conferences on Neural Networks, Spain, June (Best Student Presentation Award).

Siek, M.B. and Solomatine, D. P. (2010). *Predicting Storm Surges: Multi-models, Computational Intelligence, Chaos, Uncertainty.* Exploring Complex Dynamics in High-Dimensional Chaotic Systems: From Weather Prediction to Oceanic Flows, Dresden, Germany, January.

Siek, M.B. and Solomatine, D. P. (2009). *Chaotic Model with Data Assimilation Using NARX Network*. In Proceedings of the IEEE International Joint Conferences on Neural Networks, Atlanta, USA, June (Best Student Presentation Award).

Siek, M.B. and Solomatine, D.P. (2009). *Multivariate Phase Space Dimensionality Reduction and Chaos Model Prediction*. In Proceedings of the International Conference on Hydroinformatics, Concepción, Chile.

Siek, M.B. and Solomatine, D. P. (2008). *Multivariate Chaotic Models vs Neural Networks in Predicting Storm Surge Dynamics*. In Proceedings of the IEEE International Joint Conferences on Neural Networks, Hong Kong, June.

Siek, M.B., Solomatine, D. P. and Verlaan, M. (2008). *Short-term Prediction Storm Surges in the North Sea Using Multivariate Chaotic Model*. The EuroGOOS Conference: Coastal to Global Operational Oceanography: Achievements and Challenges, Exeter, UK, May.

Siek, M.B. and Solomatine, D. P. (2007). *Recurrence Plots in the Analysis of Extreme Ocean Storm Surges*. The 2nd International Workshop on Recurrence Plots, Siena, Italy, September.

Siek, M.B. and Solomatine, D.P. (2005). *Optimizing mixtures of local experts in tree-like regression models*. In Proceedings of the IASTED International Conference on Artificial Intelligence and Applications (AIA 2005), Innsbruck, Austria.

Solomatine, D.P. and Siek, M.B. (2004). *Semi-optimal hierarchical regression models and ANNs*. In Proceedings of the IEEE International Joint Conferences on Neural Networks, Budapest, Hungary.

Solomatine, D.P. and Siek, M.B. (2004). *Flexible and optimal M5 model trees with applications to flow predictions*. In Proceedings of the 6th International Conference on Hydroinformatics, Liong, Phoon & Babovic (eds), World Scientific Press, Singapore.

In abstracts at the international conferences

Siek, M.B. and Solomatine, D. P. (2010). *Building Chaotic Model from Incomplete Time Series*. Geophysical Research Abstracts, Vienna, Austria, May (EGU Young Scientist's Travel Award).

Siek, M.B. and Solomatine, D. P. (2009). *Performance Comparison of the European Storm Surge Models and Chaotic Model in Prediction Extreme Storm Surges*. Geophysical Research Abstracts, Vienna, Austria, April.

Siek, M.B. and Solomatine, D. P. (2009). *Dimensionality Reduction for Multivariate Phase Space Reconstruction*. Geophysical Research Abstracts, Vienna, Austria, April.

Siek, M.B. and Solomatine, D. P. (2008). *Multivariate Phase Space Reconstruction and Chaos Model Prediction for Water level and Surge Dynamics.* Geophysical Research Abstracts, Vienna, Austria, April.

Gedurende de afgelopen eeuwen hebben wereldwijd zeer ernstige kustoverstromingen plaatsgevonden vanwege stormvloeden met vaak verwoestende gevolgen. De fysische processen die leiden tot kustoverstromingen zijn inmiddels goed bekend. De ernst van een stormvloed hangt vooral af van meteorologische krachten, zoals luchtdruk verschil, windsnelheid en windrichting. De meteorologische omstandigheden worden beïnvloed door de snelheid van de depressiesystemen die zich over zee verplaatsen. Wanneer de wind het water richting de kust stuwt, kan deze uitgroeien tot wat wordt aangeduid als een stormvloed. Als een bepaalde hoge opstuwing optreedt in combinatie met hoog water ten gevolge van het getij, versterken beide effecten elkaar wat kan resulteren in een verhoogde zeespiegel met veelal ernstige overstromingen in de kustgebieden.

Nauwkeurige voorspellingen van stormvloeden zijn daarom van groot belang voor veel kustgebieden. Met name in Nederland, omdat grote delen van het land onder de zeespiegel liggen en stormvloeden vaak voorkomen op de Noordzee. De verdediging tegen overstromingen vanuit zee zijn voortdurend verbeterd, zoals door de bouw van stormvloedkeringen ontworpen voor extreme condities met een verwachtingswaarde van 1/10.000 jaar, maar ook door het bouwen van geavanceerde waarschuwingsmodellen voor het voorspellen van stormvloeden. Deze voorspellingen en waarschuwingen worden opgesteld door de Nederlandse Stormvloedwaarschuwingsdienst (SVSD) van Rijkswaterstaat, in nauwe samenwerking met het Koninklijk Nederlands Meteorologisch Instituut (KNMI). Voor een goede afsluiting van de beweegbare stormvloedkeringen dienen modelvoorspellingen tenminste 6 uur van tevoren bekend te zijn. Deze voorspellingen zijn gebaseerd op een numeriek hydrodynamisch model: het Nederlands Continentaal Plat Model (DCSM) die meteorologische voorspellingen ontvangt van het hoge resolutie model voor een beperkt gebied (HiRLAM) als aandrijvende kracht. Data assimilatie technieken gebaseerd op Ensemble Kalman filtering worden toegevoegd aan dit system om de nauwkeurigheid van de voorspellingen te verbeteren door assimilatie van recente observaties van meetstations. Andere belangrijke verbeteringen die toegevoegd zijn aan het model zijn: het verfijnen van de rekenroosters, het calibreren van het model door gebruik te maken van betere numerieke rekenschema's, en het implementeren van data assimilatie technieken (3D/4D Var en Kalman filtering). Hierbij dient te worden opgemerkt dat de nauwkeurigheid van de voorspelling van een stormvloed model gebaseerd op de Navier-Stokes vergelijkingen, zoals DCSM, voornamelijk afhankelijk is van de nauwkeurigheid van de meteorologische voorspelling van het weer model (i.c. HiRLAM).

Het genoemde model behoort tot de (klasse van) proces modellen (ook wel genoemd fysisch-gebaseerd, of numerieke modellen). Het huidige onderzoek richt zich op een heel ander model paradigma, bekend als op gegevens gebaseerd (data-driven) modellen (DDM). DDM is een modelleringtechniek die vooral gebruik maakt van de analyse van kenmerkende gegevens van het onderliggende systeem. Dit model wordt voornamelijk bepaald op basis van verbanden tussen de toestanden van de systeem variabelen (input, interne variabelen en output grootheden) met slechts beperkte kennis over details van het fysieke gedrag van het systeem. De benaderingen in data-gedreven modellen zijn over het algemeen afkomstig uit statistische methoden en kunstmatige intelligentie. Verschillende populaire modellen in DDM zijn onder andere: neurale netwerken (ANNs), leren van voorbeelden, modellen op basis van beslissingsbomen, Bayesiaans leren, commissie machines, fuzzy rule based systeem en genetisch programmeren.

Nog een andere benadering bij op data gebaseerde modellen betreft het gebruik van niet-lineaire dynamica methoden en chaos theorie, welke veel toegepast worden op het modelleren van complexe dynamische systemen. Deze methoden worden effectief toegepast en uitgebreid onderzocht, sinds Edward Lorenz in 1963 een ontdekking deed tijdens een experiment met een eenvoudig atmosferisch model. Hij onderzocht de gevoeligheid van een systeem voor de beginvoorwaarden wat leidde tot de ontwikkelling van de zogenaamde chaos theorie. Dit betekent dat een dynamisch system, afgeleid van differentiaalvergelijking, chaotisch gedrag kan vertonen, wat gekenmerkt wordt door een exponentiële divergentie van de output van het model terwijl de oorspronkelijke waarden slechts heel weinig zijn verstoord. Inmiddels heeft een groot aantal onderzoekers en wetenschappers verschillende soorten natuurlijke fenomenen onderzocht en gemodelleerd waarbij ze tot de ontdekking kwamen dat ze met specifiek chaotisch gedrag te maken hadden, terwijl eerder verondersteld werd dat deze natuurlijke systemen zich volkomen willekeurig gedroegen. Dynamische systemen die een specifiek kenmerkend gedrag vertonen, ook wel genaamd deterministische chaos, zijn voorspelbaar. In dit onderzoek hadden we de luxe te beschikken over erg grote data sets die het dynamische system kenmerken, in dit geval stormvloeden, wat ons de mogelijkheid geeft aan te tonen dat het systeem chaotisch gedrag vertoont zonder direct gebruik te hoeven maken van de differentiaalvergelijkingen die dit systeem beschrijven, en waarmee een voorspellend model gebouwd kan worden, uitsluitend op data gebaseerd (data driven).

De belangrijkste doelstelling van dit onderzoek is om een nauwkeuriger chaotisch (op data gebaseerd) model te bouwen dat kan dienen als een aanvulling op de bestaande operationele stormvloed modellen voor de Noordzee regio. Meer specifieke doelstellingen zijn (i) het analyseren van de tekortkomingen van het bestaande model, (ii) het verbeteren van technieken om een voorspellend chaotisch model te bouwen, (iii) data assimilatie methoden op te nemen in de chaotische modellen, (iv) het ontwikkelen en testen van een multi-model ensemble aanpak om diverse voorspellende modellen te combineren.Belangrijkste onderdelen van de methodiek zijn niet-lineaire systeemdynamica en chaostheorie, op data gebaseerde modellen, proces gebaseerd modelleren, data-assimilatie, optimalisatie en ensemble methoden. In algemene zin valt deze studie onder het vakgebied van de hydroinformatica. De belangrijkste case study in dit onderzoek betreft het voorspellen van de waterstanden bij het getijdestation van Hoek van Holland voor de Noord Zee. Ook hebben we aantal benaderingen getest voor de optimalisatie van chaotische modellen die gebruik maken van de data van de waterstanden bij San Juan getijde station (Puerto Rico) in de Caribische Zee.

De eerste experimenten om een uni-variabel chaotisch model te bouwen om stormvloeden te voorspellen op de Noord Zee is gedaan door Solomatine et. al. (2000). In de PhD dissertatie van Velickov (2004) werd deze benadering verder uitgewerkt en werd het voorspellende chaotische model (PCM) een multi-variabel model, waar ook andere variabelen zoals wind en luchtdruk werden opgenomen. Bij de niet-lineaire tijdreeksanalyse van de waargenomen stijging op basis van gegevens blijkt dat de stormvloed dynamiek langs de Nederlandse kust kan worden gekarakteriseerd als deterministische chaos. Chaotisch gedrag in de storm surge dynamiek kan te wijten zijn aan het feit dat dit dynamisch systeem het resultaat is van complexe interacties tussen verschillende krachten of dynamische systemen, zoals: atmosferische dynamica, wind-golf-getijde interacties, etc. De aanwezigheid van deterministische chaos met een grote positieve Lyapunov exponent impliceert de mogelijkheid om te voorspellen. Echter, de voorspelbaarheid van elk voorspellend chaotisch model heeft een aantal beperkingen. Eigenschappen als gevoeligheid voor de initiële conditie en het bestaan van bifurcaties in de oplossing kunnen redenen zijn die in verband gebracht kunnen worden met de exponentieel afnemende juistheid van de voorspelling van het chaotische model afhankelijk van de voorspellingshorizon. Toch zijn de korte en middellange termijn voorspellingen van zo'n model over het algemeen vrij nauwkeurig.

Voor het opstellen van een voorspellend chaotisch model is het noodzakelijk de waargenomen tijdsreeks van een dynamisch systeem te reconstrueren en op te nemen in een adequate m-dimensionale fase-ruimte met tijdvertragende coördinaten. Deze reconstructie behoudt de eigenschappen van het dynamische system, dat niet verandert bij een soepele aanpassing van de coördinaten, maar het behoudt niet de geometrische structuurvorm in de fase-ruimte. De juiste tijdvertragingswaarde en ingesloten dimensie kunnen geschat worden door middel van verschillende niet-lineaire analytische modellen (bijv. de eerste minimale wederzijdse informatie en correlatie dimensie,), of optimalisatie methoden. Gezien de juiste dimensie en tijdvertraging van een fase ruimte, zal de attractor van een dynamisch systeem moeten worden bepaald om vervolgens een glad overgangstraject te kunnen verkrijgen. Voorspellingen in een chaotisch model kunnen op twee manier gedaan worden: door het gebruik van globale en lokale modellen.

Bij globale modellen wordt het gehele dynamische gedrag van het system, zoals beschreven in de fase-ruimte, gekarakteriseerd en voorspeld door één globaal model. Daarentegen worden lokale modeller gekenmerkt door het dynamische gedrag ter plaatse door middel van een aantal locale modellen te bepalen, wat meer flexibiliteit biedt.De lokale modellen worden gebouwd op basis van de dynamische buren, gevonden in de fase-ruimte. Een aantal beschikbare op data gebaseerde technieken (dat wil zeggen: lineaire of niet-lineaire regressie methoden zoals (ANN) kunnen worden gebruikt als lokale modellen. Niettemin vormt de flexibiliteit van de lokale modellen een uitdaging om de beste zoektechniek te selecteren die de juiste dynamische buren en het geschikte aantal dynamische buren kan selecteren dat gebruikt kan worden voor de bouw van de voorspellende lokale modellen. De juiste buren verwijst hier naar de buren die soortgelijke dynamische kenmerken of eigenschappen (d.w.z. de zelfde soort stormontwikkeling) hebben m.b.t. concrete punten in de fase-ruimte. In dit onderzoek wordt gebruikt gemaakt van de Euclidische afstand methode om deze dynamische buren te vinden. Het eerdere gebruikte zoekalgoritme bleek niet erg selectief en vond soms dynamische buren die niet gelijkwaardige dynamische kernmerken hadden, waardoor zij ten onrechte gebruikt werden als buren bij het algoritme. In dit onderzoek werd hier speciale aandacht aan besteed, en een nieuwe zoektechniek, de zgn. traject gebaseerde methode, geïntroduceerd om onechte buren te vermijden.

De methoden en ook een aantal software componenten uit eerder onderzoek zijn geïntegreerd, getest op nieuwe data en op sommige gebieden aanzienlijk verbeterd. Dit onderzoek heft de volgende innovaties teweeg heeft gebracht.

Er is een nieuw algoritme ontwikkeld en getest om de juiste buren te kunnen identificeren. Deze zgn. traject gebaseerde methode komt voort uit het idee dat het vinden van juiste buren niet alleen afhankelijk is van de afstand tussen twee punten in de m-dimensionale fase-ruimte, maar ook van de afstand tussen de twee verschillende trajecten (sequenties van punten in de faseruimte), die deels gevormd zijn door deze twee punten. De buren worden verkregen door te zoeken naar het traject dat het dichts in de buurt is en in dezelfde richting gaat als het werkelijke traject (een traject gevormd door referentie- of het werkelijke punt in de fase-ruimte). Andere manieren om onjuiste buren te vermijden worden ook in dit onderzoek voorgesteld, zoals het gebruik van multi-step voorspellingstechniek en de afstand cut-off methode.

Het identificeren van de juiste embedding dimensies is één van de meest besproken onderwerpen binnen de gemeenschap van de niet-lineaire dynamica en chaostheorie. Zo is bijvoorbeeld een correlatie dimensie een veel gebruikte methode voor het schatten van de embedding dimensie. Deze schatting vraagt om een grote hoeveelheid tijdreeksen om een goede embedding dimensie te kunnen vaststellen. In dit onderzoek wordt het resultaat van de correlatie dimensie vergeleken met de (on)juiste naaste buren, met Cao's methode, Kaplan-Yorke of Lyapunov dimensies en met de prestaties van de optimalisatie. Computationele intelligentie, zoals raster zoeken, genetische algoritme (GA) en adaptieve cluster dekkende optimalisatie (ACCO) worden in dit onderzoek gebruikt voor het optimaal functioneren.

Ook zijn verschillende andere innovatieve ontwikkelingen gebruikt in het voorspellende chaotische model, inclusief fase-ruimte dimensionaliteit reductie, het bouwen van een chaotisch model van onvolledige tijdreeksen en het corrigeren van fase voorspellingsfouten. De niet-lineaire analyse van de tijdreeksen van een dynamisch systeem kan wijzen op de hoogdimensionale faseruimte wederopbouw. Een principaal component analyse (PCA) techniek is gebruikt om de fase-ruimte dimensie te verminderen naar een lagere dimensionale fase-ruimte met behoud van belangrijke informatie (principale componenten) van een hoge dimensionale fase-ruimte (d.w.z. informatie op afstand).De toepassing van PCA heeft nog een ander voordeel hier,namelijk het verwijderen van ruis dat kan ontstaan in de gegevens. Omdat meetinstrumenten en data-transmissie niet altijd feilloos werken in de praktijk, is het van het grootste belang een procedure in te bouwen op basis van de onvolledige tijdreeks. De mogelijkheid dat data verloren kan gaan moet ook worden onderkend bij het bouwen van een model. Om dit probleem op te lossen worden verschillende algoritmen zoals de gewogen som van lineaire interpolatie, de Bayesiaanse PCD en de "cubic spline' interpolatie voorgesteld. Een aanpak om een model te bouwen om de fase afwijking dynamisch te karakteriseren wordt voorgesteld voor het corrigeren van fase voorspellingsfouten in het chaotische model. Twee soorten modellen worden gebruikt als afwijkende voorspellers (voorspellend chaotisch model en ANN), die in staat zijn het dynamische gedrag van de afwijkende fase te identificeren en te voorspellen op basis van een standaard chaotisch model.

Een aantal methoden wordt getest om de problemen in verband met de gevoeligheid voor beginvoorwaarden en de beperking van de voorspelbaarheid van ieder model aan te pakken, inclusief een voorspellend chaotisch model. Het probleem is niet op te lossen door de gevoeligheid van de begin voorwaarde op de precieze en exacte initiële condities te bepalen. Door een data-assimilatie schema aan te brengen in het chaotische voorspellend model bestaat de mogelijkheid dit op te lossen. Een niet-lineair autoregressief neuraal netwerk met exogene ingangen (Nonlinear AutoRegressive with eXogenous inputs – NARX) is geïmplementeerd als een bijna real-time data-assimilatie techniek voor het assimileren van nieuwe geobserveerde gegevens in het voorspellend chaotische model. Deze techniek kan de lage nauwkeurigheid van de voorspellingen na een bepaalde tijd effectief corrigeren, wat vervolgens leidt tot de uitbreiding van de voorspelbaarheid van het chaotische model.

Een andere innovatie is het gebruik van multi-model ensemble voorspelling wat gezien kan worden als een effectieve manier om de voorspellingsprestatie (op basis van bias-variantie decompositie) te verbeteren. Het is vaak de moeite waard om een combinatie van verschillende voorspellingsmodellen te zoeken in plaats van alleen de beste te selecteren, zeker als die slechts marginaal de beste zou kunnen zijn. Voor het combineren van de heterogene vormen van voorspellende chaotische modellen worden multi-model ensemble voorspelling met behulp van dynamische gemiddelden en een dynamisch neurale netwerk geïntroduceerd. Een dynamische gemiddelde methode is hier geïntroduceerd - een combinatie van model selectie en combinatie van methoden gebaseerd op de prestaties van het model over een bepaalde tijd van voorspellingen. Een andere techniek maakt gebruik van dynamische neurale netwerken, een zgn. gericht tijdsvertragend neuraal netwerk (Focused Time-Delayed Neural Network – FTDNN). Verschillende voorspellingen van diverse types voorspellende chaotische modellen worden geselecteerd en gecombineerd door deze twee technieken om zo meer accurate en betrouwbare voorspellingen te verkrijgen. Met betrekking tot een hoogdimensionaal chaotisch systeem betekent dit dat een ensemble van alle toekomstige trajecten in de fase-ruime wordt geschat op basis van individuele heterogene modellen.

Een aantal verbeterde methoden om voorspellende chaotische modellen te bouwen is uitgevoerd en getest. De resultaten toonden een verhoogde voorspelling en prestatie ten opzicht van het initiële voorspellende chaotische model: PCM is 63% nauwkeuriger dan het ANN model; univariabel-PCM met PCA kan de nauwkeurigheid met 118% verhogen in vergelijking tot multi-variabele ANN; het gebruik van een PCM afwijking corrector kan de prestatie verhogen met 94%; verminderde nauwkeurigheid van -8% is het geval bij cubic spline interpolatie wanneer 30% van de waarden niet aanwezig is, de traject gebaseerde methode is beter in het vinden van de juiste buren wat resulteert in een verbetering van 185%; ACCO bleek de meest efficiënte optimisatie techniek te zijn voor het voorspellend chaotisch model, wat leidde tot een toename van 67% van de nauwkeurigheid; data assimilatie met gebruik van een NARX netwerk bracht 553% verbetering; multi-model ensemble voorspellingen die gebruik maken van FTDNN met 'batch learning' bleek de meest effectieve methode om de prestatie van het voorspellend chaotisch model te verhogen met 967%. Toch, zal er nog verder onderzoek gedaan moeten worden om de betrouwbaarheid van de verbeterde methoden en de mogelijkheden om deze te combinerenen, te testen.

Over het geheel genomen draagt dit onderzoek bij aan de ontwikkeling om extreme waterstanden beter te voorspellen. De modeltechnieken gebaseerd op de methoden van niet-lineaire dynamica, chaostheorie, statistiek en neurale netwerken met diverse verbeteringen en innovaties hebben aangetoond dat het voorspellende chaotische model kan dienen als een efficiënt hulpmiddel voor accurate en betrouwbare korte termijn voorspellingen van stormvloeden ter ondersteuning van beleidsmakers bij het nemen van beslissingen in geval van overstromingen en ten behoeve van navigatie. We geloven dat deze aanpak goede mogelijkheden biedt om als aanvullende methode te worden gebruikt in de praktijk, samen met de traditionele numerieke oceaanmodellen.

Delft, 6 December 2011

Michael Siek

T - #0119 - 071024 - C238 - 280/208/13 - PB - 9780415621021 - Gloss Lamination